Statistical Physics

Statistical Physics

Second Revised and Enlarged Edition

by

Tony Guénault

Emeritus Professor of Low Temperature Physics
Lancaster University, UK

A C.I.P. Catalogue record for this book is available from the Library of Congress.

ISBN 978-1-4020-5974-2 ISBN 978-1-4020-5975-9 (eBook)
DOI 10.1007/978-1-4020-5975-9

www.springer.com

Printed on acid-free paper

First edition 1988
Second edition 1995
Reprinted 1996, 2000, 2001, 2003
Reprinted revised and enlarged second edition 2007

Table of contents

Preface

Preface to the first edition

Statistical physics is not a difficult subject, and I trust that this will not be found a difficult book. It contains much that a number of generations of Lancaster students have studied with me, as part of their physics honours degree work. The lecture course was of 20 hours' duration, and I have added comparatively little to the lecture syllabus. A prerequisite is that the reader should have a working knowledge of basic thermal physics (i.e. the laws of thermodynamics and their application to simple substances). The book *Thermal Physics* by Colin Finn in this series forms an ideal introduction.

Statistical physics has a thousand and one different ways of approaching the same basic results. I have chosen a rather down-to-earth and unsophisticated approach, without I hope totally obscuring the considerable interest of the fundamentals. This enables applications to be introduced at an early stage in the book.

As a low-temperature physicist, I have always found a particular interest in statistical physics, and especially in how the absolute zero is approached. I should not, therefore, apologize for the low-temperature bias in the topics which I have selected from the many possibilities.

Without burdening them with any responsibility for my competence, I would like to acknowledge how much I have learned in very different ways from my first three 'bosses' as a trainee physicist: Brian Pippard, Keith MacDonald and Sydney Dugdale. More recently my colleagues at Lancaster, George Pickett, David Meredith, Peter McClintock, Arthur Clegg and many others have done much to keep me on the rails. Finally, but most of all, I thank my wife Joan for her encouragement.

A.M. Guénault
1988

Preface to the second edition

Some new material has been added to this second edition, whilst leaving the organization of the rest of the book (Chapters 1–12) unchanged. The new chapters aim to illustrate the basic ideas in three rather distinct and (almost) independent ways. Chapter 13 gives a discussion of chemical thermodynamics, including something about chemical equilibrium. Chapter 14 explores how some interacting systems can still be treated by a simple statistical approach, and Chapter 15 looks at two interesting applications of statistical physics, namely superfluids and astrophysics.

The book will, I hope, be useful for university courses of various lengths and types. Several examples follow:

1. Basic general course for physics undergraduates (20–25 lectures): most of Chapters 1–12, omitting any of Chapters 7, 10, 11 and 12 if time is short;
2. Short introductory course on statistical ideas (about 10 lectures): Chapters 1, 2 and 3 possibly with material added from Chapters 10 and 11;
3. Following (2), a further short course on statistics of gases (15 lectures): Chapters 4–6 and 8–9, with additional material available from Chapter 14 and 15.2;
4. For chemical physics (20 lectures): Chapters 1–7 and 10–13;
5. As an introduction to condensed matter physics (20 lectures): Chapters 1–6, 8–12, 14, 15.1.

In addition to those already acknowledged earlier, I would like to thank Keith Wigmore for his thorough reading of the first edition and Terry Sloan for his considerable input to my understanding of the material in section 15.2.1.

A.M. Guénault
1994

Preface to the revised and enlarged second edition

This third edition of Statistical Physics follows the organization and purpose of the second edition, with comparatively minor updating and changes to the text. I hope it continues to provide an accessible introduction to the subject, particularly suitable for physics undergraduates. Chapter summaries have been added to the first nine (basic) chapters, in order to encourage students to revise the important ideas of each chapter – essential background for an informed understanding of later chapters.

A.M. Guénault
2007

A SURVIVAL GUIDE TO STATISTICAL PHYSICS

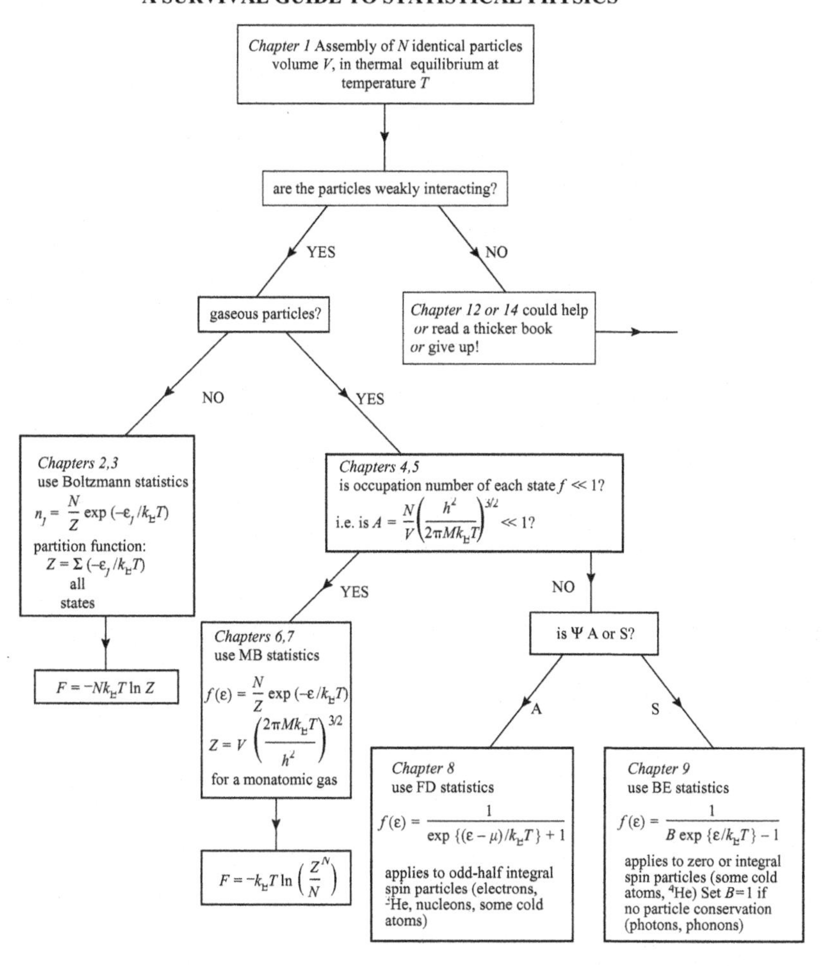

1

Basic ideas

There is an obvious problem about getting to grips with an understanding of matter in thermal equilibrium. Let us suppose you are interested (as a designer of saucepans?) in the thermal capacity of copper at 450 K. On the one hand you can turn to thermodynamics, but this approach is of such generality that it is often difficult to see the point. Relationships between the principal heat capacities, the thermal expansion coefficient and the compressibility are all very well, but they do not help you to understand the particular magnitude and temperature dependence of the actual heat capacity of copper. On the other hand, you can see that what is needed is a microscopic mechanical picture of what is going on inside the copper. However, this picture becomes impossibly detailed when one starts to discuss the laws of motion of 10^{24} or so copper atoms.

The aim of statistical physics is to make a bridge between the over-elaborate detail of mechanics and the obscure generalities of thermodynamics. In this chapter we shall look at one way of making such a bridge. Most readers will already be familiar with the kinetic theory of ideal gases. The treatment given here will enable us to discuss a much wider variety of matter than this, although there will nevertheless be some limitations to the traffic that can travel across the bridge.

1.1 THE MACROSTATE

The basic task of statistical physics is to take a system which is in a well-defined thermodynamic state and to compute the various thermodynamic properties of that system from an (assumed) microscopic model.

The '*macrostate*' is another word for the thermodynamic state of the system. It is a specification of a system which contains just enough information for its thermodynamic state to be well defined, but no more information than that. As outlined in most books on thermal physics (e.g. Finn's book *Thermal Physics* in this series), for the simple case of a pure substance this will involve:

- the nature of the substance – e.g. natural copper;
- the amount of the substance – e.g. 1.5 moles;

- a small number of pairs of thermodynamic co-ordinates – e.g. pressure P and volume V; magnetic field B and magnetization M; surface tension and surface area, etc.

Each of these pairs is associated with a way of doing work on the system. For many systems only $P - V$ work is relevant, and (merely for brevity) we shall phrase what follows in terms of $P - V$ work only. Magnetic systems will also appear later in the book.

In practice the two co-ordinates specified, rather than being P and V, will be those appropriate to the external conditions. For instance, the lump of copper might be at a specific pressure P (= 1 atm) and temperature T (= 450 K). In this case the macrostate would be defined by P and T; and the volume V and internal energy U and other parameters would then all be determined in principle from P and T. It is precisely one of the objectives of statistical physics to obtain from first principles what are these values of V, U, etc. (In fact, we need not set our sights as low as this. Statistical physics also gives detailed insights into dynamical properties, and an example of this is given in Chapter 12.)

Now comes, by choice, an important limitation. In order to have a concrete situation to discuss in this chapter (and indeed throughout the first eight chapters of this book), we shall concentrate on one particular type of macrostate, *namely that appropriate to an isolated system*. Therefore the macrostate will be defined by the nature of the substance, the amount, and by U and V. For the isolated system in its energy-proof enclosure, the internal energy is a fixed constant, and V is also constant since no work is to be done on the system. The (fixed) amount of the substance we can characterize by the number N of microscopic 'particles' making up the system.

This limitation is not too severe in practice. For an isolated system in which N is reasonably large, fluctuations in (say) T are small and one finds that T is determined really rather precisely by (N, U, V). Consequently one can use results based on the (N, U, V) macrostate in order to discuss equally well the behaviour in any other macrostate, such as the (N, P, T) macrostate appropriate to our piece of copper.

Towards the end of the book (Chapters 12 and 13, in particular), we shall return to the question as to how to set up methods of statistical physics which correspond to other macrostates.

1.2 MICROSTATES

Let us now consider the mechanical microscopic properties of the system of interest, which we are assuming to be an assembly of N identical microscopic particles. For the given (N, U, V) macrostate there are an enormous number of possible '*microstates*'.

The word microstate means the most detailed specification of the assembly that can be imagined. For example, in the classical kinetic theory of gases, the microstate would need to specify the (vector) position and momentum of each of the N gas particles, a total of $6N$ co-ordinates. (Actually even this is assuming that each particle is a structureless point, with no internal degrees of freedom like rotation, vibration, etc.) Of course, this microstate contains a totally indigestible amount of information, far too much for one to store even one microstate in the largest available computer. But, worse still, the system changes its microstate very rapidly indeed – for instance one mole of a typical gas will change its microstate roughly 10^{32} times a second.

Clearly some sort of averaging over microstates is needed. And here is one of those happy occasions where quantum mechanics turns out to be a lot easier than classical mechanics.

The conceptual problem for classical microstates, as outlined above for a gas, is that they are infinite in number. The triumph of Boltzmann in the late 19th century – had he lived to see the full justification of it – and of Gibbs around the turn of the century, was to see how to do the averaging nevertheless. They observed that a system spends equal times in equal volumes of 'phase-space' (a combined position and momentum space; we shall develop these ideas much later in the book, in section 14.4). Hence the volume in phase-space can be used as a statistical weight for microstates within that volume. Splitting the whole of phase-space into small volume elements, therefore, leads to a feasible procedure for averaging over all microstates as required. However, we can nowadays adopt a much simpler approach.

In quantum mechanics a microstate by definition is *a quantum state of the whole assembly*. It can be described by a single N-particle wave function, containing all the information possible about the state of the system. The point to appreciate is that quantum states are discrete in principle. Hence although the macrostate (N, U, V) has an enormous number of possible microstates consistent with it, the number is none the less definite and finite. We shall call this number Ω, and it turns out to play a central role in the statistical treatment.

1.3 THE AVERAGING POSTULATE

We now come to the assumption which is the whole basis of statistical physics:

All accessible microstates are equally probable.

This averaging postulate is to be treated as an assumption, but it is of interest to observe that it is nevertheless a reasonable one. Two types of supporting argument can be produced.

The first argument is to talk about time-averages. Making any physical measurement (say, of the pressure of a gas on the wall of its container) takes a non-zero time; and in the time of the measurement the system will have passed through a *very* large number of microstates. In fact this is why we get a reproducible value of P; observable fluctuations are small over the appropriate time scale. Hence it is reasonable that we should be averaging effectively over all accessible microstates. The qualification 'accessible' is included to allow for the possibility of metastability. There can be situations in which groups of microstates are not in fact accessed in the time scale of the measurement, so that there is in effect another constant of the motion, besides N, U and V; only a subset of the total number Ω of microstates should then be averaged. We shall return to this point in later chapters, but will assume for the present that all Ω microstates are readily accessible from each other. Hence the time-average argument indicates that averaging over all microstates is necessary. The necessity to average *equally* over all of them is not so obvious, rather it is assumed. (In passing one can note that for a gas this point relates to the even coverage of classical phase-space as mentioned above, in that quantum states are evenly dispersed through phase-space; for example see Chapter 4.)

The second type of supporting argument is to treat the postulate as a 'confession of ignorance', a common stratagem in quantum mechanics. Since we do not in fact know which one of the Ω microstates the system is in at the time of interest, we simply average equally over all possibilities, i.e. over all microstates. This is often called an 'ensemble' average, in that one can visualize it as replicating the measurement in a whole set of identical systems and then averaging over the whole set (or ensemble).

One can note that the equality of ensemble and time averages implies a particular kind of uniformity in a thermodynamic system. To give an allied social example, consider the insurer's problem. He wishes to charge a fair (sic) premium for life insurance. Thus he requires an expectation of life for those currently alive, but he cannot get this by following them with a stop-watch until they die. Rather, he can look at biographical records in the mortuary in order to determine an expectation of life (for the wrong sample) and hope for uniformity.

1.4 DISTRIBUTIONS

In attempting to average over all Ω microstates we still have a formidable problem. A typical system (e.g. a mole of gas) is an assembly of $N = 10^{24}$ particles. That is a large enough number, but the number Ω of microstates is of order N^N, an astronomically large number. We must confess that knowledge of the system at the microstate level is too detailed for us to handle, and therefore we should restrict our curiosity merely to a *distribution* specification, defined below.

A distribution involves assigning individual (private) energies to each of the N particles. This is only sensible (or indeed possible) for an assembly of *weakly interacting particles*. The reason is that we shall wish to express the total internal energy U of the assembly as the sum of the individual energies of the N particles

$$U = \sum_{l=1}^{N} \varepsilon(l) \tag{1.1}$$

where $\varepsilon(l)$ is the energy of the lth particle. Any such expression implies that the interaction energies between particles are much smaller than these (self) energies ε.

Actually any thermodynamic system must have some interaction between its particles, otherwise it would never reach equilibrium. The requirement rather is for the interaction to be small enough for (1.1) to be valid, hence 'weakly interacting' rather than 'non-interacting' particles.

Of course this restriction of approach is extremely limiting, although less so than one might first suspect. Clearly, since the restriction is also one of the assumptions of simple kinetic theory, our treatment will be useful for perfect gases. However, it means that for a real fluid having strong interactions between molecules, i.e. an imperfect gas or a liquid, the method cannot be applied. We shall return briefly to this point in Chapter 14, but a full treatment of interacting particles is well outside the scope of this book. At first sight it might seem that a description of solids is also outside this framework, since interactions between atoms are obviously strong in a solid. However, we shall see that many of the thermal properties of solids are nevertheless to be understood from a model based on an assembly of N weakly interacting particles, when one recognizes that these particles need not be the atoms, but other appropriate entities. For example the particles can be phonons for a discussion of lattice vibrations (Chapter 9); localized spins for a discussion of magnetic properties (Chapters 2 and 11); or conduction electrons for a description of metals and semiconductors (Chapter 8).

A *distribution* then relates to the energies of a single particle. For each microstate of the assembly of N identical weakly interacting particles, each particle is in an identifiable one-particle state. In the distribution specification, intermediate in detail between the macrostate and a microstate, we choose not to investigate *which* particles are in which states, but only to specify the total *number* of particles in the states.

We shall use two alternative definitions of a distribution.

Definition 1 – Distribution in states This is a set of numbers $(n_1, n_2, \ldots, n_j, \ldots)$ where the typical distribution number n_j is defined as the number of particles in state j, which has energy ε_j. Often, but not always, this distribution will be an infinite set; the label j must run over all the possible states for one particle. A useful shorthand for the whole set of distribution numbers $(n_1, n_2, \ldots, n_j, \ldots)$ is simply $\{n_j\}$.

The above definition is the one we shall adopt until we specifically discuss gases (Chapter 4 onwards), at which stage an alternative, and somewhat less detailed, definition becomes useful.

Definition 2 – Distribution in levels This is a set of numbers $(n_1, n_2, \ldots, n_i, \ldots)$ for which the typical number n_i is now defined as the number of particles in level i, which has energy ε_i and degeneracy g_i, the degeneracy being defined as the number of states belonging to that level. The shorthand $\{n_i\}$ will be adopted for this distribution.

It is worth pointing out that the definition to be adopted is a matter of one's choice. The first definition is the more detailed, and is perfectly capable of handling the case of degenerate levels – degeneracy simply means that not all the ε_js are different. We shall reserve the label j for the states description and the label i for the levels description; it is arguable that the n symbols should also be differentiated, but we shall not do this.

Specifications – an example Before proceeding with physics, an all too familiar example helps to clarify the difference between the three types of specification of a system, the macrostate, the distribution, and the microstate.

The example concerns the marks of a class of students. The macrostate specification is that the class of 51 students had an average mark of 55%. (No detail at all, but that's thermodynamics.) The microstate is quite unambiguous and clear; it will specify the name of each of the 51 individuals and his/her mark. (Full detail, nowhere to hide!) The definition of the distribution, as above, is to some extent a matter of choice. But a typical distribution would give the number of students achieving marks in each decade, a total of 10 distribution numbers. (Again all identity of individuals is lost, but more statistical detail is retained than in the macrostate.)

1.5 THE STATISTICAL METHOD IN OUTLINE

The object of the exercise is now to use the fundamental averaging assumption about microstates (section 1.3) to discover the particular distribution $\{n_j\}$ (section 1.4) which best describes the thermal equilibrium properties of the system.

We are considering an isolated system consisting of a fixed number N of the identical weakly interacting particles contained in a fixed volume V and with a fixed internal energy U. There are essentially four steps towards the statistical description of this macrostate which we discuss in turn:

I. solve the one-particle problem;
II. enumerate possible distributions;
III. count the microstates corresponding to each distribution;
IV. find the average distribution.

1.5.1 The one-particle problem

This is a purely mechanical problem, and since it involves only one particle it is a soluble problem for many cases of interest. The solution gives the states of a particle which we label by j (= 0, 1, 2, ...). The corresponding energies are ε_j. We should note that these energies depend on V (for a gas) or on V/N the volume per particle (for a solid).

1.5.2 Possible distributions

The possible sets of distribution numbers $\{n_j\}$ can now be simply written down (given appropriate patience, because usually there will be very many possibilities). However, we give this relatively straightforward task a section of its own, in order to stress that a valid distribution must satisfy the two conditions implied by the macrostate

$$\sum_j n_j = N \tag{1.2}$$

$$\sum_j n_j \varepsilon_j = U \tag{1.3}$$

Equation (1.2) ensures that the distribution contains the correct number of particles. Equation (1.3) follows from (1.1), and guarantees that distribution corresponds to the correct value of U. All of the conditions of the (N, U, V) macrostate are now taken care of.

1.5.3 Counting microstates

Next we need to count up the number of microstates consistent with each valid set of distribution numbers. Usually, and especially for a large system, each distribution $\{n_j\}$ will be associated with a very large number of microstates. This number we call $t(\{n_j\})$. The dependence of t on $\{n_j\}$ is a pure combinatorial problem. The result is very different for an assembly of localized particles (in which the particles are distinguishable by their locality) and for an assembly of gas-like particles (in which the particles are fundamentally indistinguishable). Hence the statistical details for localized particles and for gases are treated below in separate chapters.

1.5.4 The average distribution

The reason for counting microstates is that, according to the postulate of equal probability of all microstates, the number $t(\{n_j\})$ is the *statistical weight* of the distribution $\{n_j\}$. Hence we can now in principle make the correct weighted average over all possible distributions to determine the average distribution $\{n_j\}_{av}$. And this average distribution, according to our postulates, is the one which describes the thermal equilibrium distribution.

1.6 A MODEL EXAMPLE

Before further discussion of the properties of a large system, the realistic case in thermodynamics, let us investigate the properties of a small model system using the methodology of the previous section.

1.6.1 A simple assembly

The macrostate we consider is an assembly of $N = 4$ distinguishable particles. We label the four particles A, B, C and D. The total energy $U = 4\varepsilon$, where ε is a constant (whose value depends on V).

Step I. The mechanical problem is to be solved to give the possible states of one particle. We take the solution to be states of (non-degenerate) energies $0, \varepsilon, 2\varepsilon, 3\varepsilon, \ldots$ For convenience we label these states $j = 0, 1, 2, \ldots$ with $\varepsilon_j = j\varepsilon$.

Step II. Defining the distribution numbers as $\{n_j\}$, with $j = 0, 1, 2, \ldots,$ we note that any allowable distributions must satisfy

$$\sum n_j = 4; \quad \sum n_j \varepsilon_j = 4\varepsilon \quad (\text{i.e.} \sum n_j j = 4)$$

There are thus five possible distributions:

Distribution	n_0	n_1	n_2	n_3	n_4	n_5	$\ldots$
1	3	0	0	0	1	0	$\ldots$
2	2	1	0	1	0	0	$\ldots$
3	2	0	2	0	0	0	$\ldots$
4	1	2	1	0	0	0	$\ldots$
5	0	4	0	0	0	0	$\ldots$

Step III. A microstate specifies the state of each of the four particles. We need to count the number of microstates to each of the five distributions. To take distribution 1 as an example, we can identify four possible microstates:

(i) A is in state $j = 4$; B, C and D are in state $j = 0$
(ii) B is in state $j = 4$, the others are in state $j = 0$
(iii) C is in state $j = 4$, the others are in state $j = 0$
(iv) D is in state $j = 4$, the others are in state $j = 0$

Hence $t^{(1)} = 4$. Similarly one can work out (an exercise for the reader?) the numbers of microstates for the other four distributions. The answers are $t^{(2)} = 12$, $t^{(3)} = 6$, $t^{(4)} = 12$, $t^{(5)} = 1$. It is significant that the distributions which spread the particles between the states as much as possible have the most microstates, and thus (Step IV) are the most probable. The total number of microstates Ω is equal to the sum $t^{(1)} + t^{(2)} + t^{(3)} + t^{(4)} + t^{(5)}$, i.e. $\Omega = 35$ in this example.

(*Aside.* A general formula for this situation is developed in Appendix A and we shall make much use of it in Chapter 2. The result is

$$t(\{n_j\}) = N! \Big/ \prod_j n_j! \tag{1.4}$$

where the denominator $\prod n_j!$ represents the extended product $n_0!n_1!n_2!\ldots n_j!\ldots$. The result for $t^{(1)}$ follows from (1.4) as $4!/3!$, when one substitutes $0! = 1$ and $1! = 1$. The other t values follow similarly.)

Step IV. The average value of every distribution number can now be obtained by an equal averaging over every microstate, readily computed as a weighted mean over the five possible distributions using the t values as the weight. For instance

$$\begin{aligned}(n_0)_{av} &= (n_0^{(1)}t^{(1)} + n_0^{(2)}t^{(2)} + \cdots)/\Omega \\ &= (3 \times 4 + 2 \times 12 + 2 \times 6 + 1 \times 12 + 0 \times 1)/35 \\ &= 60/35 = 1.71\end{aligned}$$

Similarly for n_1, n_2 etc. give finally

$$\{n_j\}_{av} = (1.71, 1.14, 0.69, 0.34, 0.11, 0, 0 \ldots)$$

The result is not unlike a falling exponential curve, the general result for a large assembly which we shall derive in the next chapter.

1.6.2 A composite assembly

We now briefly reconsider the four-particle model assembly as a composite two-part assembly. We take as an initial macrostate a more restrictive situation than before. Let us suppose that the two particles A and B form one sub-assembly, which is thermally isolated from a second sub-assembly consisting of particles C and D; and that the values of internal energies are $U_{AB} = 4\varepsilon$, $U_{CD} = 0$. This initial macrostate is more fully specified than the situation in section 1.6.1, since the division of energy between AB and CD is fixed.

The total number of microstates is now only five rather than 35. This arises as follows. For sub-assembly CD we must have $j = 0$ for both particles, the only way to achieve $U_{CD} = 0$. Hence $\Omega_{CD} = 1$ and $\{n_j\} = (2, 0, 0, 0, 0, \ldots)$. (This result corresponds to a low-temperature distribution, as we shall see later, in that all the particles are in the lowest state). For sub-assembly AB there are five microstates to give $U_{AB} = 4\varepsilon$. In a notation [j(A), j(B)] they are $[4, 0]$, $[0, 4]$, $[3, 1]$, $[1, 3]$, $[2, 2]$. Hence $\Omega_{AB} = 5$ and $\{n_j\} = (0.4, 0.4, 0.4, 0.4, 0.4, 0, 0, \ldots)$. (This is now a high-temperature distribution, with the states becoming more evenly populated.)

In order to find the total number of microstates of any composite two-part assembly, the values of Ω for the two sub-assemblies must be multiplied together. This is because for every microstate of the one sub-assembly, the other sub-assembly can be in any one of its microstates. Therefore in this case we have $\Omega = \Omega_{AB} \cdot \Omega_{CD} = 5$, as stated earlier.

Now let us place the two sub-assemblies in thermal contact with each other, while still remaining isolated from the rest of the universe. This removes the restriction of the 4:0 energy division between AB and CD, and the macrostate reverts to exactly the same as in section 1.6.1. When equilibrium is reached, this then implies a distribution intermediate in shape (and in temperature). But of particular importance is the new value of Ω (namely 35). One way of understanding the great increase (and indeed another way of calculating it) is to appreciate that microstates with $U_{AB} = 3\varepsilon, 2\varepsilon, \varepsilon$, 0 are now accessible in addition to 4ε; and correspondingly for U_{CD}. So we can observe from this example that internal adjustments of an isolated system towards overall equilibrium *increase* the number of accessible microstates.

1.7 STATISTICAL ENTROPY AND MICROSTATES

First, a word about entropy. Entropy is introduced in thermodynamics as that rather shadowy function of the state of a system associated with the second law of thermodynamics. The essence of the idea concerns the direction (the 'arrow of time') of spontaneous or natural processes, i.e. processes which are found to occur in practice.

A pretty example is the mixing of sugar and sand. Start with a dish containing a discrete pile of each substance. The sugar and sand may then be mixed by stirring, but the inverse process of re-separating the substances by un-stirring (stirring in the opposite direction?) does not in practice happen. Such un-mixing can only be achieved with great ingenuity. In a natural process, the second law tells us that the entropy S of the universe (or of any isolated system) never decreases. And in the mixing process the entropy increases. (The 'great ingenuity' would involve a larger increase of entropy somewhere else in the universe.) All this is a statement of probability, rather than of necessity – it is *possible* in principle to separate the mixed sugar and sand by stirring, but it is almost infinitely *improbable*. And thermodynamics is the science of the probable!

Statistical physics enables one to discuss entropy in terms of probability in a direct and simple way. We shall adopt in this book a statistical definition for the entropy of an isolated system

$$S = k_B \ln \Omega \tag{1.5}$$

with k_B equal to Boltzmann's constant $1.38 \times 10^{-23}\,\mathrm{J\,K^{-1}}$. This definition has an old history, originating from Boltzmann's work on the kinetic theory of gases in the last century, and (1.5) appears on Boltzmann's tombstone. The relationship was developed further by Planck in his studies of heat radiation – the start of the quantum theory.

Logically perhaps (1.5) is a derived *result* of statistical physics. However, it is such a central ideal that it is sensible to introduce it at this stage, and to treat it as a *definition* of entropy from the outset. We shall gradually see that the S so defined has all the properties of the usual thermodynamic entropy.

What we have observed so far about the behaviour of Ω is certainly consistent with this relation to entropy.

1. As noted above, for an isolated system a natural process, i.e. one which spontaneously occurs as the system attains overall equilibrium, is precisely one in which the thermodynamic entropy increases. And we have seen in the example of section 1.6 that Ω also increases in this type of process. Hence a direct relation

between S and Ω is suggested, and moreover a monotonically increasing one, in agreement with (1.5).

2. For a composite assembly, made up of two sub-assemblies 1 and 2 say, we have shown in section 1.6.2 that the number of microstates of the whole assembly Ω is given by $\Omega = \Omega_1 \cdot \Omega_2$. The required behaviour of the thermodynamic entropy is of course $S = S_1 + S_2$, and the relation (1.5) is consistent with this; indeed the logarithm is the only function which will give the result. (This was Planck's original 'proof' of (1.5).)
3. The correlation between entropy and the number of microstates accessible (i.e. essentially a measure of disorder) is a very appealing one. It interprets the third law of thermodynamics to suggest that all matter which comes to equilibrium will *order* at the absolute zero in the sense that only one microstate will be accessed ($\Omega = 1$ corresponding to $S = 0$, a natural zero for entropy).

Later in the book, we shall see much more evidence in favour of (1.5), the final test being that the results derived using it are correct, for example the equation of state of an ideal gas (Chapter 6), and the relation of entropy to temperature and heat (Chapter 2).

1.8 SUMMARY

In this chapter, the main ideas of a statistical approach to understanding thermal properties are introduced. These include:

1. Statistical methods are needed as a bridge between thermodynamics (too general) and mechanics (too detailed).
2. This bridge is readily accessible if we restrict ourselves to a system which can be considered as an assembly of weakly-interacting particles.
3. Three ways of specifying such a system are used. The macrostate corresponds to the thermodynamic specification, based on a few external constraints. The microstate is a full mechanical description, giving all possible knowledge of its internal configuration. Between these is the statistical notion of a distribution of particles which gives more detail than the macrostate, but less than the microstate.
4. The number Ω of microstates which describe a given macrostate plays a central role. The basic assumption is that all (accessible) microstates are equally probable.
5. If we define entropy as $S = k_B \ln \Omega$, then this is a good start in our quest to "understand" thermodynamics.

2

Distinguishable particles

The next step is to apply the statistical method outlined in Chapter 1 to realistic thermodynamic systems. This means addressing the properties of an assembly which consists of a large number N of weakly interacting identical particles. There are two types of assembly which fulfil the requirements.

One type is a gaseous assembly, in which the identical particles are the gas molecules themselves. In quantum mechanics one recognizes that the molecules are not only identical, but they are also (in principle as well as in practice) indistinguishable. It is not possible to 'put a blob of red paint' on one particular molecule and to follow its history. Hence the microstate description must take full account of the indistinguishability of the particles. Gaseous assemblies will be introduced later in Chapter 4.

In this chapter we shall treat the other type of assembly, in which the particles are distinguishable. The physical example is that of a solid rather than that of a gas. Consider a simple solid which is made up of N identical atoms. It remains true that the atoms themselves are indistinguishable. However, a good description of our assembly is to think about the solid as a set of N lattice sites, in which each lattice site contains an atom. A 'particle' of the assembly then becomes 'the atom at lattice site 4357 (or whatever)'. (*Which* of the atoms is at this site is not specified.) The particle is distinguished not by the identity of the atom, but by the distinct location of each lattice site. A solid is an assembly of localized particles, and it is this locality which makes the particles distinguishable.

We shall now develop the statistical description of an ideal solid, in which the atoms are weakly interacting. How far the behaviour of real solids can be explained in this way will become clearer in later chapters. The main results of this chapter will be the derivation of the thermal equilibrium distribution (the Boltzmann distribution) together with methods for the calculation of thermodynamic quantities. Two physical examples are given in Chapter 3.

2.1 THE THERMAL EQUILIBRIUM DISTRIBUTION

We follow the method outlined in section 1.5. For the macrostate we consider an assembly of N identical distinguishable (localized) particles contained in a fixed volume V and having a fixed internal energy U. The system is mechanically and thermally isolated, but we shall be considering sufficiently large assemblies that the other thermodynamic quantities (T, S, etc.) are well defined.

2.1.1 The one-particle states

The one-particle states will be specified by a state label j $(= 0, 1, 2 \ldots)$. The corresponding energies ε_j may or may not be all different. These states will be dependent upon the volume per particle (V/N) for our localized assembly.

2.1.2 Possible distributions

We use the distribution in states $\{n_j\}$ defined in section 1.4. The distribution numbers must satisfy the two conditions (1.2) and (1.3) imposed by the macrostate

$$\sum_j n_j = N \qquad \text{(1.2) and (2.1)}$$

$$\sum_j n_j \varepsilon_j = U \qquad \text{(1.3) and (2.2)}$$

2.1.3 Counting microstates

It is here that the fact that we have distinguishable particles shows up. It means that the particles can be counted just as macroscopic objects. A microstate will specify the state (i.e. the label j) for each distinct particle. We wish to count up how many such micro-states there are to an allowable distribution $\{n_j\}$. The problem is essentially the same as that discussed in Appendix A, namely the possible arrangements of N objects into piles with n_j in a typical pile. The answer is

$$t(\{n_j\}) = N! \Big/ \prod_j n_j! \qquad \text{(1.4) and (2.3)}$$

2.1.4 The average distribution

According to our postulate, the thermal distribution should now be obtained by evaluating the average distribution $\{n_j\}_{\text{av}}$. This task involves a weighted average of all the possible distributions (as allowed by (2.1) and (2.2)) using the values of t

(equation (2.3)) as statistical weights. This task can be performed, but fortunately we are saved from the necessity of having to do anything so complicated by the large numbers involved. Some of the simplifications of large numbers are explored briefly in Appendix B.

The vital point is that it turns out that one distribution, say $\{n_j^*\}$, is overwhelmingly more probable than any of the others. In other words, the function $t(\{n_j\})$ is very sharply peaked indeed around $\{n_j^*\}$. Hence, rather than averaging over all possible distributions, one can obtain essentially the same result by picking out the most probable distribution alone. This then reduces to the mathematical problem of maximizing $t(\{n_j\})$ from (2.3) subject to the conditions (2.1) and (2.2).

Another even stronger way of looking at the sharp peaking of t is to consider the relationship between Ω and t. Since Ω is defined as the total number of microstates contained by the macrostate, it follows that

$$\Omega = \sum t(\{n_j\})$$

where the sum goes over all distributions. What is now suggested (and compare the pennies problem of Appendix B) is that this sum can in practice be replaced by its maximum term, i.e.

$$\Omega \approx t(\{n_j^*\}) = t^* \text{(for short)} \tag{2.4}$$

2.1.5 The most probable distribution

To find the thermal equilibrium distribution, we need to find the maximum t^* and to identify the distribution $\{n_j^*\}$ at this maximum. Actually it is a lot simpler to work with $\ln t$, rather than t itself.

Since $\ln x$ is a monotonically increasing function of x, this does not change the problem; it just makes the solution a lot more straightforward. Taking logarithms of (2.3) we obtain

$$\ln t = \ln N! - \sum_j \ln n_j! \tag{2.5}$$

Here the large numbers come to our aid. Assuming that all the ns are large enough for Stirling's approximation (Appendix B) to be used, we can eliminate the factorials to obtain

$$\ln t = (N \ln N - N) - \sum_j (n_j \ln n_j - n_j) \tag{2.6}$$

To find the maximum value of $\ln t$ from (2.6) we express changes in the distribution numbers as differentials (large numbers again!) so that the maximum will be obtained by differentiating $\ln t$ and setting the result equal to zero. Using the fact that N is constant, and noting the cancellation of two out of the three terms arising from the sum in (2.6), this gives simply

$$\begin{aligned} \mathrm{d}(\ln t) &= 0 - \sum_j \mathrm{d}n_j(\ln n_j + n_j/n_j - 1) \\ &= -\sum_j \ln n_j^* \mathrm{d}n_j = 0 \end{aligned} \tag{2.7}$$

where the $\mathrm{d}n_j$s represent any allowable changes in the distribution numbers from the required distribution $\{n_j^*\}$. Of course not all changes are allowed. Only changes which maintain the correct values of N (2.1) and of U (2.2) may be countenanced. This lack of independence of the $\mathrm{d}n_j$s gives two restrictive conditions, obtained by differentiating (2.1) and (2.2)

$$\mathrm{d}(N) = \sum_j \mathrm{d}n_j = 0 \tag{2.8}$$

$$\mathrm{d}(U) = \sum_j \varepsilon_j \mathrm{d}n_j = 0 \tag{2.9}$$

A convenient way of dealing with the mathematics of a restricted maximum of this type is to use the Lagrange method of undetermined multipliers. The argument in our case goes as follows. First we note that we can self-evidently add any arbitrary multiples of (2.8) and (2.9) to (2.7), and still achieve the result zero. Thus

$$\sum_j (-\ln n_j^* + \alpha + \beta\varepsilon_j)\mathrm{d}n_j = 0 \tag{2.10}$$

for *any* values of the constants α and β. The second and clever step is then to recognize that it will always be possible to write the solution in such a way that each individual term in the sum of (2.10) equals zero, so long as specific values of α and β are chosen. In other words the most probable distribution $\{n_j^*\}$ will be given by

$$(-\ln n_j^* + \alpha + \beta\varepsilon_j) = 0 \tag{2.11}$$

with α and β each having a specific (but as yet undetermined) value. This equation can then be written

$$n_j^* = \exp(\alpha + \beta\varepsilon_j) \tag{2.12}$$

and this is the *Boltzmann distribution*. We may note at once that it has the exponential form suggested for the thermal equilibrium distribution by the little example in Chapter 1. But before we can appreciate the significance of this central result, we need to explore the meanings of these constants α and β. (*A note of caution:* α and β

are defined with the opposite sign in some other works. Looking at a familiar result (e.g. the Boltzmann distribution) easily makes this ambiguity clear.)

2.2 WHAT ARE α AND β?

Here (as ever?) physics and mathematics go hand in hand. In the mathematics, α was introduced as a multiplier for the number condition (2.1), and β for the energy condition (2.2). So it will follow that α is determined from the fixed number N of particles, and can be thought of as a 'potential for particle number'. Similarly β is determined by ensuring that the distribution describes an assembly with the correct energy U, and it can be interpreted as a 'potential for energy'.

2.2.1 α and number

Since α enters the Boltzmann distribution in such a simple way, this section will be short! We determine α by applying the condition (2.1) which caused its introduction in the first place. Substituting (2.12) back into (2.1) we obtain

$$N = \sum_j n_j = \exp(\alpha) \sum_j \exp(\beta\varepsilon_j) \tag{2.13}$$

since $\exp(\alpha)$ ($=A$, say) is a factor in each term of the distribution. In other words, A is a normalization constant for the distribution, chosen so that the distribution describes the thermal properties of the correct number N of particles. Another way of writing (2.13) is: $A = N/Z$, with the 'partition function', Z, defined by $Z = \sum_j \exp(\beta\varepsilon_j)$. We may then write the Boltzmann distribution (2.12) as

$$n_j = A\exp(\beta\varepsilon_j) = (N/Z)\exp(\beta\varepsilon_j) \tag{2.14}$$

We leave any fuller discussion of the partition function until later in the chapter, when the nature of β has been clarified.

2.2.2 β and energy

In contrast, the way in which β enters the Boltzmann distribution is more subtle. Nevertheless the formal statements are easily made. We substitute the thermal distribution (2.14) back into the relevant condition (2.2) to obtain

$$U = \sum_j n_j\varepsilon_j = (N/Z)\sum_j \varepsilon_j \exp(\beta\varepsilon_j)$$

or

$$U/N = \sum_j \varepsilon_j \exp(\beta\varepsilon_j) \Big/ \sum_j \exp(\beta\varepsilon_j) \tag{2.15}$$

The appropriate value of β is then that one which, when put into this equation, gives precisely the internal energy per particle (U/N) specified by the macrostate. Unfortunately this is not a very tidy result, but it is as far as we can go explicitly, since one cannot in general invert (2.15) to give an explicit formula for β as a function of (U, V, N). Nevertheless for a given (U, V, N) macrostate, β is fully specified by (2.15), and one can indeed describe it as a 'potential for energy', in that the equation gives a very direct correlation between (U/N) and β.

However, that is not the end of the story. It turns out that this untidy (but absolutely specific) function β has a very clear physical meaning in terms of thermodynamic functions other than (U, V, N). In fact we shall see that it must be related to *temperature* only, a sufficiently important point that the following section will be devoted to it.

2.3 A STATISTICAL DEFINITION OF TEMPERATURE

2.3.1 β and temperature

To show that there is a necessary relationship between β and temperature T, we consider the thermodynamic and statistical treatments of two systems in thermal equilibrium.

The thermodynamic treatment is obvious. Two systems in thermal equilibrium have, effectively by definition, the same temperature. This statement is based on the 'zeroth law of thermodynamics', which states that there is some common function of state shared by all systems in mutual thermal equilibrium – and this function of state is what is meant by (empiric) temperature.

The statistical treatment can follow directly the lines of section 2.1. The problem can be set up as follows. Consider two systems P and Q which are in thermal contact with each other, but together are isolated from the rest of the universe. The statistical method applies to this composite assembly in much the same way as in the simple example of section 1.6.2. We suppose that system P consists of a fixed number N_P of localized particles, which each have states of energy ε_j, as before. The system Q need not be of the same type, so we take it as containing N_Q particles whose energy states are ε'_k. The corresponding distributions are $\{n_j\}$ for system P and $\{n'_k\}$ for system Q. Clearly the restrictions on the distributions are

$$\sum_j n_j = N_P \tag{2.16}$$

$$\sum_k n'_k = N_Q \tag{2.17}$$

and

$$\sum_j n_j \varepsilon_j + \sum_k n'_k \varepsilon'_k = U \tag{2.18}$$

where U is the total energy (i.e. $U_P + U_Q$) of the two systems together.

The counting of microstates is easy when we recall (section 1.6.2) that we may write

$$\Omega = \Omega_P \times \Omega_Q \quad \text{or} \quad t = t_P \times t_Q \tag{2.19}$$

with the expressions for t_P and t_Q being analogous to (2.3).

Finding the most probable distribution may again be achieved by the Lagrange method as in section 2.1.5. The problem is to maximize $\ln t$ with t as given in (2.19), subject now to the three conditions (2.16), (2.17) and (2.18). Using multipliers α_P, α_Q and β respectively for the three conditions, the result on differentiation (compare (2.12)) is

$$\sum_j (-\ln n_j^* + \alpha_P + \beta\varepsilon_j)\mathrm{d}n_j + \sum_k (-\ln n_j'^* + \alpha_Q + \beta\varepsilon'_k)\mathrm{d}n_k = 0$$

The Lagrange method then enables one to see that, for the appropriate values of the three multipliers, each term in the two sums is separately equal to zero, so that the final result is for system P

$$n_j^* = \exp(\alpha_P + \beta\varepsilon_j)$$

and for system Q

$$n_k'^* = \exp(\alpha_Q + \beta\varepsilon'_k)$$

Where does this leave us? What it shows is that both system P and system Q have their thermal equilibrium distributions of the Boltzmann type (compare (2.12)). The distributions have their own private values of α, and we can see from the derivation that this followed from the introduction of the separate conditions for particle conservation ((2.16) and (2.17)). However, the two distributions have the same value of β. This arose in the derivation from the single energy condition (2.18), in other words from the thermal contact or energy interchange between the systems. So the important conclusion is that two systems in mutual thermal equilibrium and distributions with the same β. From thermodynamics we know that they necessarily have the same empiric temperature, and thus the same thermodynamic temperature T. Therefore it follows that β *is a function of* T *only*.

2.3.2 Temperature and entropy

We now come to the form of this relation β and T. The simplest approach is to know the answer(!), and we shall choose to *define* a statistical temperature in terms of β from the equation

$$\beta = -1/k_B T \tag{2.20}$$

What will eventually become clear (notably when we discuss ideal gases in Chapter 6) is that this definition of T does indeed agree with the absolute thermodynamic scale of temperature. Meanwhile we shall adopt (2.20) knowing that its justification will follow.

There is much similarity with our early definition of entropy as $S = k_B \ln \Omega$, introduced in section 1.7. And in fact the connection between these two results is something well worth exploring at this stage, particularly since it can give us a microscopic picture of heat and work in reversible processes.

Consider how the internal energy U of a system can be changed. From a macroscopic viewpoint, this can be done by adding heat and/or work, i.e. change in U = heat input + work input. The laws of thermodynamics for a differential change in a simple $P - V$ system tell us that

$$\mathrm{d}U = T\mathrm{d}S - P\mathrm{d}V \tag{2.21}$$

where for reversible processes (only) the first ($T\mathrm{d}S$) term can be identified as the heat input, and the second term ($-P\mathrm{d}V$) as the work input.

Now let us consider the microscopic picture. The internal energy is simply the sum of energies of all the particles of the system, i.e. $U = \sum n_j \varepsilon_j$. Taking again a differential change in U, we obtain

$$\mathrm{d}U = \sum_j \varepsilon_j \mathrm{d}n_j + \sum_j n_j \mathrm{d}\varepsilon_j \tag{2.22}$$

where the first term allows for changes in the occupation numbers n_j, and the second term for changes in the energy levels ε_j.

It is not hard to convince oneself that the respective first and second terms of (2.21) and (2.22) match up. The energy levels are dependent only on V, so that $-P\mathrm{d}V$ work input can only address the second term of (2.22). And, bearing in mind the correlation between S and Ω (and hence t^*, and hence $\{n_j^*\}$), it is equally clear that occupation number changes are directly related to entropy changes. Hence the matching up of the first terms. These ideas turn out to be both interesting and useful. The relation $-P\mathrm{d}V = \sum n_j \mathrm{d}\varepsilon_j$ gives a direct and physical way of calculating the pressure from a microscopic model. And the other relation bears directly on the topic of this section.

The argument in outline is as follows. Start by considering how a change in $\ln \Omega$ can be brought about

$$
\begin{aligned}
\mathrm{d}(\ln \Omega) &= \mathrm{d}(\ln t^*) && \text{for a large system} \\
&= -\sum_j \ln n_j^* \mathrm{d}n_j && \text{as in (2.7)} \\
&= -\sum_j (\alpha + \beta\varepsilon_j)\mathrm{d}n_j && \text{using the Boltzmann distribution} \\
&= -\beta \sum_j \varepsilon_j \mathrm{d}n_j && \text{since } N \text{ is fixed} \\
&= -\beta(\mathrm{d}U)_{\text{no work}} && \text{first term of (2.22)} \\
&= -\beta(T\mathrm{d}S) && \text{first term of (2.21)}
\end{aligned}
$$

This identification is consistent with $S = k_B \ln \Omega$ together with $\beta = -1/k_B T$. It shows clearly that the two statistical definitions are linked, i.e. that (2.20) validates (1.5) or vice versa. Part of this consistency is to note that it must be the same constant (k_B, Boltzmann's constant) which appears in both definitions.

2.4 THE BOLTZMANN DISTRIBUTION AND THE PARTITION FUNCTION

We now return to discuss the Boltzmann distribution. We have seen that this distribution is the appropriate one to describe the thermal equilibrium properties of an assembly of N identical localized (distinguishable) weakly interacting particles. We have derived it for an isolated assembly having a fixed volume V and a fixed internal energy U. An important part in the result is played by the parameter β which is a function of the macrostate (U, V, N). However, the upshot of the previous section is to note that the Boltzmann distribution is most easily written and understood in terms of (T, V, N) rather than (U, V, N). This is no inconvenience, since it frequently happens in practice that it is T rather than U that is known. And it is no embarrassment from a fundamental point of view so long as we are dealing with a large enough system that fluctuations are unimportant. Therefore, although our method logically determines T (and other thermodynamic quantities) as a function of (U, V, N) for an isolated system, we shall usually use the results to describe the behaviour of U (and other thermodynamic quantities) as a function of (T, V, N). (This subject will be reopened in Chapters 10–12.)

Therefore, we now write the Boltzmann distribution as

$$n_j = (N/Z)\exp(-\varepsilon_j/k_B T) \tag{2.23}$$

with the partition function Z defined as

$$Z = \sum_j \exp(-\varepsilon_j/k_B T) \tag{2.24}$$

(From now on we shall for simplicity omit the *s from n_j, since all the discussion will relate to thermal equilibrium.) It is worth making two points about the partition function. The first is that its common symbol Z is used from the German word for sum-over-states, for that is all the partition function is: the sum over all one-particle states of the 'Boltzmann factors' $\exp(-\varepsilon_j/k_B T)$. The second point concerns its English name. It is called the partition function because (in thermal equilibrium at temperature T) n_j is proportional to the corresponding term in the sum. In other words the N particles are partitioned into their possible states (labelled by j) in just the same ratios as Z is split up into the Boltzmann factor terms. This is clear when we rewrite (2.23) as

$$n_j/N = \exp(-\varepsilon_j/k_B T)/Z \tag{2.25}$$

or equivalently

$$n_j/n_k = \exp[-(\varepsilon_j - \varepsilon_k)/k_B T] \tag{2.26}$$

In fact the way of writing the Boltzmann distribution given in (2.26) is a very straight-forward way of remembering it. And expressions of the type $\exp(-\Delta\varepsilon/k_B T)$ turn up in all sorts of different physical situations.

2.5 CALCULATION OF THERMODYNAMIC FUNCTIONS

To finish this chapter we discuss a few practicalities about how the Boltzmann distribution may be used to calculate thermodynamic functions from first principles. In the next chapter we apply these ideas to two particular physical cases.

We start with our system at given (T, V, N) as discussed in the previous section. There are then (at least) three useful routes for calculation. The best one to use will depend on which thermodynamic functions are to be calculated – and there is no substitute for experience in deciding!

Method 1: Use $S = k_B \ln \Omega$ This method is often the shortest to use if only the entropy is required. The point is that one can substitute the Boltzmann distribution numbers, (2.23), back into (2.3) in order to give t^* and hence Ω (equation (2.4)) and hence S. Thus S is obtained from a knowledge of the ε_js (which depend on V), of N and of T (as it enters the Boltzmann distribution).

Method 2: Use the definition of Z There is a direct shortcut from the partition function to U. This is particularly useful if only U and perhaps dU/dT (= C_V, the heat capacity at constant volume) are wanted. In fact U can be worked out at once from $U = \sum n_j \varepsilon_j$, but this sum can be neatly performed by looking back at (2.15), which can be re-expressed as

$$(U/N) = (1/Z)\mathrm{d}Z/\mathrm{d}\beta = \mathrm{d}(\ln Z)/\mathrm{d}\beta \tag{2.27}$$

Note that it is usually convenient to retain β as the variable here rather than to use T.

Method 3: The 'royal route' This one never fails, so it is worth knowing! The route uses the Helmholtz free energy F, defined as $F = U - TS$. The reason for the importance of the method is twofold. First, the statistical calculation of F turns out to be extremely simple from the Boltzmann distribution. The calculation goes as follows

$$\begin{aligned}
F &= U - TS && \text{definition}\\
&= \sum n_j \varepsilon_j - k_B T \ln t^* && \text{statistical } U \text{ and } S\\
&= \sum n_j \varepsilon_j - k_B T (N \ln N - \sum n_j \ln n_j) && \text{using (2.6) with } \sum n_j = N\\
&= -N k_B T \ln Z, \text{ simply} && \text{using (2.23)}
\end{aligned} \tag{2.28}$$

In the last step one takes the logarithm of (2.23) to obtain $\ln n_j = \ln N - \ln Z - \varepsilon_j / k_B T$. Everything but the $\ln Z$ term cancels, giving the memorable and simple result (2.28).

The second reason for following this route is that an expression for F in terms of (T, V, N) is of immediate use in thermodynamics since (T, V, N) are the natural co-ordinates for F (e.g. see *Thermal Physics* by Finn, Chapter 10). In fact $\mathrm{d}F = -S\mathrm{d}T - P\mathrm{d}V + \mu \mathrm{d}N$, so that simple differentiation can give S, P and the chemical potential μ at once; and most other quantities can also be derived with little effort. We shall see how these ideas work out in the next chapter.

2.6 SUMMARY

This chapter lays the groundwork for the statistical method which is developed in later chapters.

1. We first consider an assembly of *distinguishable* particles, which makes counting of microstates a straightforward operation. This corresponds to several important physical situations, two of which follow in Chapter 3.
2. Counting the microstates leads to (2.3), a result worth knowing.
3. The statistics of large numbers ensures that we can accurately approximate the average distribution by the most probable.
4. Use of 'undetermined multipliers' demonstrates that the resulting Boltzmann distribution has the form $n_j = \exp(\alpha + \beta \varepsilon_j)$.
5. α relates to the number N of particles, leading to the definition of the partition function, Z.

6. β is a 'potential for energy U' and thus relates to temperature. Note the inconsistency in the sign of β between different authors as a possible cause of confusion.
7. The formula $\beta = -1/k_{\mathrm{B}}T$ gives a statistical definition of temperature, which agrees with the usual thermodynamic definition.
8. Inclusion of T explicitly in the Boltzmann distribution is often useful in applications in which we specify a (T, V, N) macrostate.

3

Two examples

We now apply the general results of the previous chapter to two specific examples, chosen because they are easily soluble mathematically, whilst yet being of direct relevance to real physical systems. The first example is an assembly whose localized particles have just two possible states. In the second the particles are harmonic oscillators.

3.1 A SPIN-$\frac{1}{2}$ SOLID

First we derive the statistical physics of an assembly whose particles have just two states. Then we apply the results to an ideal 'spin-$\frac{1}{2}$ solid'. Finally we can examine briefly how far this model describes the observed properties of real substances, particularly in the realm of ultra-low temperature physics.

3.1.1 An assembly of particles with two states

Consider an assembly of N localized weakly interacting particles in which there are just two one-particle states. We label these states $j = 0$ and $j = 1$, with energies (under given conditions) ε_0 and ε_1. The results of the previous chapter can be used to write down expressions for the properties of the assembly at given temperature T.

The distribution numbers The partition function Z, (2.24), has only two terms. It is $Z = \exp(-\varepsilon_0/k_B T) + \exp(-\varepsilon_1/k_B T)$, which can be conveniently written as

$$Z = \exp(-\varepsilon_0/k_B T)[1 + \exp(-\varepsilon/k_B T)] \quad (3.1)$$

where the energy ε is defined as the difference between the energy levels $(\varepsilon_1 - \varepsilon_0)$. We may note that (3.1) may be written as $Z = Z(0) \times Z(\text{th})$, the product of two factors. The first factor ($Z(0) = \exp(-\varepsilon_0/k_B T)$, the so-called zero-point term) depends only on the ground-state energy ε_0, whereas the second factor ($Z(\text{th})$, the thermal term) depends only on the relative energy ε between the two levels.

The thermal equilibrium distribution numbers (or occupation numbers) n_0 and n_1 may now be evaluated from (2.23) to give

$$n_0 = N/[1 + \exp(-\varepsilon/k_B T)] = N/Z(\text{th})$$

and

$$\begin{aligned} n_1 &= N \exp(-\varepsilon/k_B T)/[1 + \exp(-\varepsilon/k_B T)] \\ &= N \exp(-\varepsilon/k_B T)/Z(\text{th}) \end{aligned} \tag{3.2}$$

These numbers are sketched in Fig. 3.1 as functions of T. We can note the following points

1. The numbers $n_{0,1}$ do not depend at all on the ground state energy ε_0. The first factor $Z(0)$ of (3.1) does not enter into (3.2). Instead all the relevant information is contained in $Z(\text{th})$. This should cause no surprise since all is contained in the simple statement (2.26) of the Boltzmann distribution, namely $n_1/n_0 = \exp(-\varepsilon/k_B T)$, together with $n_0 + n_1 = N$. Hence also:
2. The numbers $n_{0,1}$ are functions only of $\varepsilon/k_B T$. One can think of this as the ratio of two 'energy scales'. One scale ε represents the separation between the two energy levels of the particles, and is determined by the applied conditions, e.g. of volume V (or for our spin-$\frac{1}{2}$ solid of applied magnetic field – see later). The second energy scale is $k_B T$, which should be thought of as *a thermal energy scale*. Equivalently, the variable may be written as θ/T, the ratio of two temperatures, where $\theta = \varepsilon/k_B$ is a temperature characteristic of the energy-level spacing. This idea of temperature or energy scales turns out to be valuable in many other situations also.
3. At low temperatures (meaning $T \ll \theta$, or equivalently $k_B T \ll \varepsilon$), equation (3.2) gives $n_0 = N, n_1 = 0$. To employ useful picture language, all the particles are

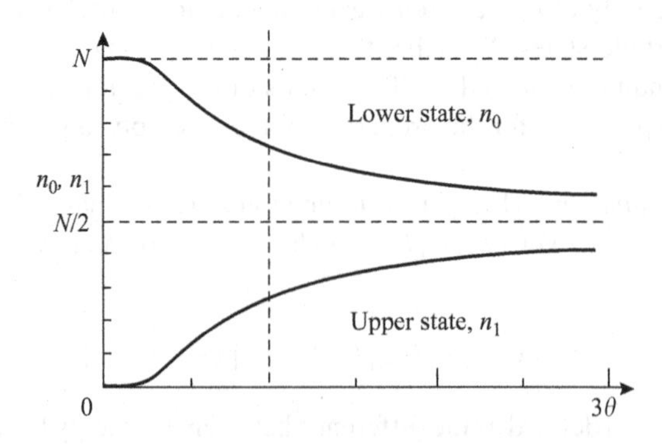

Fig. 3.1 Occupation numbers for the two states of a spin-$\frac{1}{2}$ solid in thermal equilibrium at temperature T. The characteristic temperature θ depends only on the energy difference between the two states. The particles are all in the lower state when $T \ll \theta$, but the occupation of the two states becomes equal when $T \gg \theta$.

frozen out into the lowest energy (ground) state, and no particle is excited into the higher state.

4. On the other hand at high temperatures (meaning $T \gg \theta$ or $k_B T \gg \varepsilon$) we have $n_0 = n_1 = N/2$. There are equal numbers in the two states, i.e. the difference in energy between them has become an irrelevance. The probability of any particle being in either of the two states is the same, just like in the penny-tossing problem of Appendix B.

Internal energy U An expression for U can be written down at once as

$$\begin{aligned} U &= n_0 \varepsilon_0 + n_1 \varepsilon_1 \\ &= N\varepsilon_0 + n_1 \varepsilon \end{aligned} \tag{3.3}$$

with n_1 given by (3.2). This function is sketched in Fig. 3.2. The first term in (3.3) is the 'zero-point' energy, $U(0)$, the energy $T = 0$. The second term is the 'thermal energy', $U(\text{th})$, which depends on the energy level spacing ε and $k_B T$ only.

One should note that this expression may also be obtained directly from the partition function (3.1) using (2.27). The $Z(0)$ factor leads to $U(0)$ and the $Z(\text{th})$ factor to the $U(\text{th})$. It is seen from Fig. 3.2 that the transition from low to high temperature behaviour again occurs around the characteristic temperature θ.

Heat capacity C The heat capacity C (strictly C_V, or in general the heat capacity at constant energy levels) is obtained by differentiating U with respect to T. The zero-point term goes out, and one obtains

$$C = Nk_B \frac{(\theta/T)^2 \exp(-\theta/T)}{[1 + \exp(-\theta/T)]^2} \tag{3.4}$$

The result, plotted in Fig. 3.3, shows a substantial maximum (of order Nk_B, the natural unit for C) at a temperature near to θ. C vanishes rapidly (as $\exp(-\varepsilon/k_B T)/T^2$)

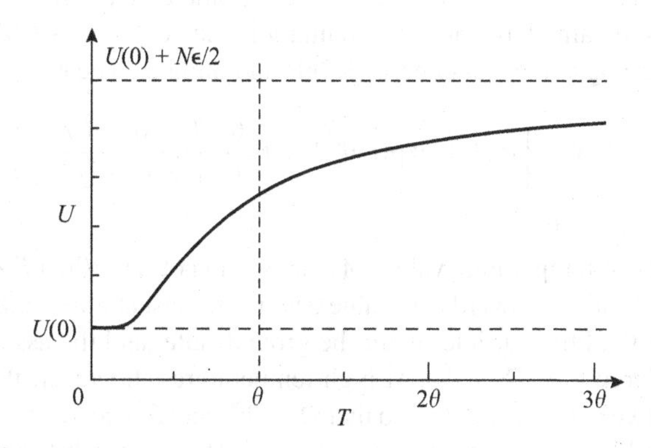

Fig. 3.2 The variation of internal energy with temperature for a spin-$\frac{1}{2}$ solid. $U(0)$ is the zero-point energy.

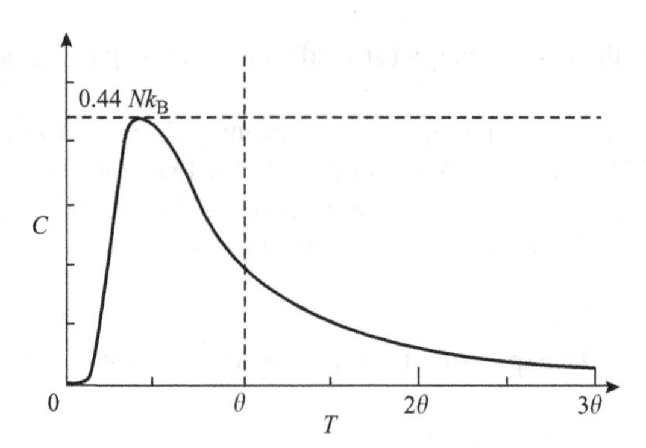

Fig. 3.3 The variation of heat capacity with temperature for a spin-$\frac{1}{2}$ solid, showing the peak of the Schottky anomaly (see section 3.1.2).

at low temperatures. It also goes to zero, albeit rather more gently as $1/T^2$, at high temperatures. (Prove this last statement? Seen by expanding the exponentials in (3.4).)

Entropy S Finally we obtain the entropy S. Again this is a thermal property only, independent of the zero-point term. With section 2.5 before us we have three possible derivations. Method 1 can be used (together with Stirling's approximation) to give the answer from the distribution numbers (3.2) without recourse to calculus. Alternatively (method 2) one can obtain S from an integration of C up from the absolute zero (since $S = 0$ there), i.e. $S = \int_0 (C/T)\mathrm{d}T$. This is possible but not recommended. Instead let us use method 3. From (2.28) we write down

$$\begin{aligned} F &= -Nk_B T \ln Z \\ &= N\varepsilon_0 - Nk_B T \ln[1 + \exp(-\varepsilon/k_B T)] \end{aligned} \tag{3.5}$$

substituting (3.1) for Z. Again note the zero-point and thermal terms in (3.5). The entropy is obtained by one differentiation, since $S = -(\partial F/\partial T)_{V,N}$, the differentiation being at constant energy levels and number. The result

$$S = Nk_B \left\{ \ln[1 + \exp(-\theta/T)] + \frac{(\theta/T)\exp(-\theta/T)}{[1 + \exp(-\theta/T)]} \right\} \tag{3.6}$$

is illustrated in Fig. 3.4.

The high and low temperature values of S are worth noting. When $T \ll \theta$, S is again frozen out exponentially towards the value 0 which it must attain eventually (the third law!). As $T \to 0$ all the particles enter the ground state, and the assembly becomes completely ordered (i.e. $\Omega = 1$). At high temperatures, however, the particles are randomized between the two states, so that $\Omega = 2^N$, and S reaches the expected value $Nk_B \ln 2$ (again like the penny-tossing problem). The approach to the high T limit again goes as $1/T^2$.

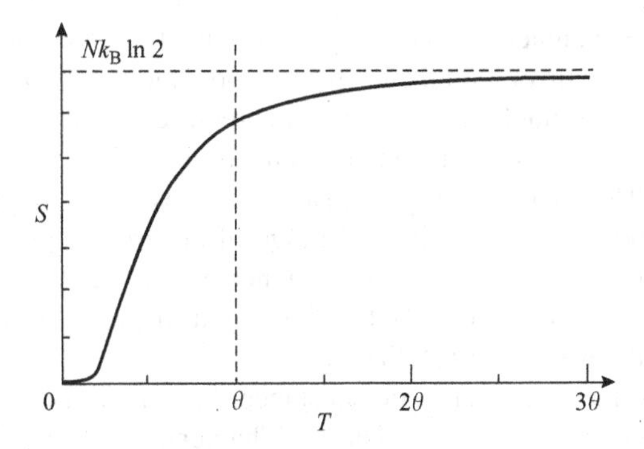

Fig. 3.4 The variation of entropy with temperature for a spin-$\frac{1}{2}$ solid. Note the transition from perfect order ($S = 0$) towards total disorder ($S = Nk_B \ln 2$).

3.1.2 Magnetic solids

Actually the two-state situation is quite a common one. It occurs because of the existence of spin states in atoms. If a single atom has an angular momentum quantum number of $\frac{1}{2}$, then when it experiences a magnetic field B, the atomic energy levels split into two. One level goes up by μB and the other goes down by μB, where μ is the z-component of the magnetic moment associated with the angular momentum. This is the Zeeman effect (much loved by spectroscopists, since this splitting of $2\mu B$ can be observed by a variety of techniques). In the lowest (ground) state one can imagine the atomic moment to be aligned parallel to the field, whereas in the upper (unstable) state it is antiparallel.

There are two origins of these magnetic moments in solids. One is from the electrons in the atom, and the other from the nucleus. An electronic effect can come both from the orbital motion and also from the intrinsic spin of the electron. The precise value of μ depends on the details of the atomic origin of the moment, and need not concern us in this discussion, except to note that in order of magnitude it will equal the *Bohr magneton* $\mu_B (= e\hbar/2m = 0.93 \times 10^{-23}\,\mathrm{J\,T^{-1}})$. For simplicity we shall in future refer to the angular momentum as 'spin', regardless of its actual orbital or intrinsic origin. The second cause is nuclear spin. Here even more the magnitude of μ depends on the specific nucleus, but the much smaller order of magnitude is the *nuclear magneton* $\mu_N (= e\hbar/2M_P = 5.05 \times 10^{-27}\mathrm{J\,T^{-1}})$.

We now consider the thermal properties of a solid whose atoms have spin $\frac{1}{2}$. To whatever properties the solid would exhibit without the $2\mu B$ splitting, we must add a spin contribution. And in many situations this contribution is precisely described by the two-state model of the previous section. In order for our model to apply, all we require is that the localized spins are weakly interacting and identical. This is a good approximation if (but only if) the spins are sufficiently far apart that the B-field experienced by each spin arises almost entirely from an externally applied field, rather than from the influence of its neighbours. In the usual terminology this means that the substance is paramagnetic (alignment of permanent moments in an applied field) rather than ferromagnetic or antiferromagnetic (spontaneous alignment of permanent moments). Another profitable way of stating the requirement is that the energy levels ε_0 and ε_1 do not themselves depend on the occupation numbers n_0 and n_1, but only on the applied field. This is a good approximation for most nuclear spin systems, since μ_N is small. It applies only to a limited range of electronic spin systems, notably diluted paramagnetic salts such as cerium magnesium nitrate (CMN) which contains only a few spins (Ce^{3+}ions) separated by a lot of non-magnetic padding (all the other atoms, including the water of crystallization).

For this type of ideal paramagnetic substance the hard work is already done. (A treatment of ferromagnetism appears later, in Chapter 11.) We may use the results of section 3.1.1 to determine the contribution of the spins to the thermal properties of the solid. The energy difference ε between the two states is equal to $2\mu B$, and hence the characteristic temperature θ equals $2\mu B/k_B$. The number N is the number of spin-$\frac{1}{2}$ particles in the solid (i.e. much less than the number of atoms in the case of CMN). Note the following:

1. The thermal properties of the spins, dependent only on θ/T, are thus for a given system universal functions of B/T. We shall use this result for S in particular in the next section.
2. Typical values of θ in strong magnetic fields are a few degrees K for an electronic spin system, but a few mK for a nuclear spin system. As we shall see later in the book, there are few other thermal contributions at such low temperatures, so that the spins in fact form the major thermal bath in the solid.
3. We are restricting our discussion to a spin-$\frac{1}{2}$ solid, one which has just two spin states. The treatment for a higher spin follows very similar lines, with qualitatively identical results. The major difference is that, since a spin I has $(2I + 1)$ possible states, the high temperature entropy is $Nk_B \ln(2I + 1)$ rather than $Nk_B \ln 2$. The general form of all the functions is similar, with again a characteristic temperature θ which relates to the Zeeman splitting.
4. The heat capacity (Fig. 3.3) is worthy of note. Heat capacities of this sort are called Schottky anomalies. The word 'anomalies' is used because of the potential upset to an unsuspecting experimenter. As the range of measurement is reduced below room temperature, the large lattice contribution to a typical solid heat capacity reduces rapidly (often as T^3) towards zero. Imagine the consternation when a further reduction of T sees an *increasing* T^{-2} contribution starting to come in! However, this is precisely what happens as spins in the system become ordered as θ is reached.

3.1.3 Cooling by adiabatic demagnetization

The paramagnetic solid can form the basis of a method of cooling, and one which is of great importance in physics, since it is the only workable method in the sub-millikelvin region.

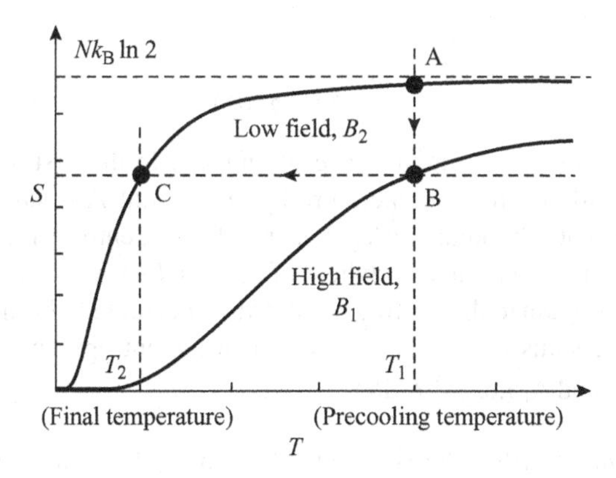

Fig. 3.5 Cooling by adiabatic magnetization. The two $S - T$ curves give the behaviour in a high magnetic field, B_1, and a low field, B_2, AB represents the precooling process (removal of entropy at a high temperature), and BC the adiabatic demagnetization to the final temperature T_2.

The theory of the method is easily followed from an $S - T$ diagram as shown in Fig. 3.5. The figure shows two entropy curves, one in the highest available field (B_1), and one in a low field (B_2). As noted above, S is a function only of B/T, so the only difference between the two curves is the scaling along the T-axis. The cooling method uses two distinct steps. Firstly the solid is magnetized at the lowest available precooling temperature, T_1. This is illustrated in Fig. 3.5 by the step AB, which represents an isothermal magnetization. In the second step, the system is then thermally isolated and the magnetic field gradually reduced, giving an adiabatic reversible (i.e. isentropic) process. In other words the system goes from B to C on the diagram, and its temperature reduces from T_1 to T_2. (Actually the spin-$\frac{1}{2}$ solid is a very suitable working substance for a refrigerator – the requirements are that S should vary with T and with another control parameter. Here the parameter is the applied magnetic field; in a more common fluid refrigerator it is the pressure.)

It is interesting to consider the two processes microscopically. In the isothermal magnetization leg (AB), the spins tend to align in the strong magnetic field since μB is comparable to $k_B T$. In this process, the population of the lower state (energy $-\mu B$, i.e. with μ parallel to B) grows at the expense of the upper state (energy $+\mu B$). The population becomes (somewhat) ordered, and the entropy decreases. During this process heat of magnetization is evolved from the system, and in fact the precooling refrigerator has to be very active to maintain the temperature at T_1. In the second adiabatic step (BC), the situation is different. The spins are no longer in contact with a heat bath, so the occupation numbers cannot change. As discussed in Chapter 2, S constant means $t(\{n_j\})$ constant which means $\{n_j\}$ constant. Hence at C the occupation numbers $n_{0,1}$ are just the same as they were at B. However, the energy level spacing has reduced (from $2\mu B_1$ to $2\mu B_2$) and so the temperature must have also decreased in the same ratio, in order to keep $\mu B/k_B T$ constant. Therefore the cooling obeys the simple law

$$T_2 = T_1(B_2/B_1) \tag{3.7}$$

This is a nice example of the importance of entropy. In the first leg we reduce the entropy of the solid at a high temperature by an amount ΔS, say, by extracting a quantity $T_1 \Delta S$ of heat. The adiabatic leg then transfers the entropy reduction to a lower temperature, and the spins can absorb heat of at least $T_2 \Delta S$ from their surroundings before warming up again to the starting point. (More accurately the amount of cooling available from the spins equals the area between the entropy curve and the S-axis between points C and A, i.e. $\int_{\mathrm{C}}^{\mathrm{A}} T \,\mathrm{d}S$).

Some experimental details We shall refer to two implementations of the cooling method. One is the use of CMN to cool to about 2 mK, and the second is the use of Cu nuclei to cool into the μK region.

The necessary starting conditions are easy to work out. To obtain a significant entropy reduction ΔS under the starting conditions, one requires μB_1 to be of the same order of magnitude as $k_B T_1$. In the case of electronic moments (CMN), this requirement is comfortably attained with a precooling temperature T_1 of around 1 K and an applied field B_1 of 1 T. For nuclear moments (Cu) the starting conditions are more stringent by a factor of over 1000. They are marginally met in modern techniques using almost the highest available magnetic field – typically a 7 T superconducting magnet, and a precooling refrigerator operating at around 10 mK. Even so only a few per cent of the copper spin entropy is removed.

The essential ingredients are indicated in Fig. 3.6. For CMN the precooling stage is usually a pumped helium cryostat (reaching about 1 K with ^{4}He, or 0.3 K with ^{3}He).

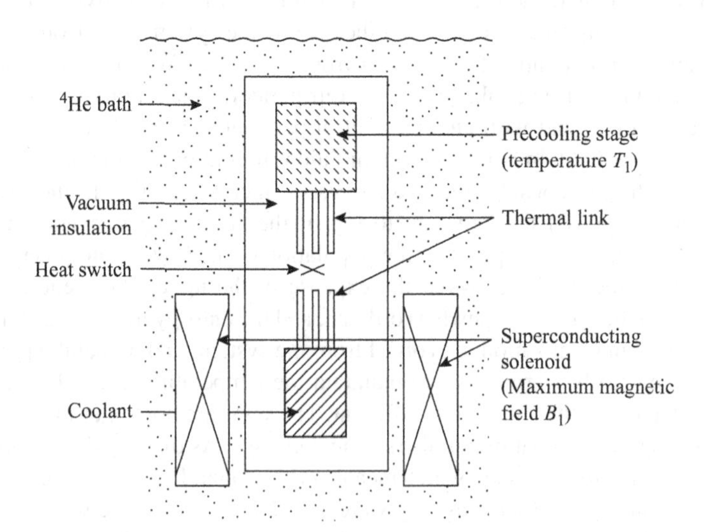

Fig. 3.6 A schematic apparatus for cooling by adiabatic demagnetization.

For nuclear spins it will be a 'dilution refrigerator', a device capable of reaching 3–10 mK whose working substance is a ^{3}He–^{4}He mixture which phase separates at temperatures below 0.7 K. The magnet nowadays is usually a superconducting solenoid operated in the main helium bath for both experiments.

The CMN coolant is often in the form of a powdered slurry around a bunch of fine copper wires for thermal contact. The active particles for nuclear cooling are the ^{63}Cu and ^{65}Cu nuclei in copper metal (both of these nuclei having spin $\frac{3}{2}$). The copper can be in the form of wires, plates or powder.

The final element in the experiment is the heat switch and thermal link. We require an arrangement which will give excellent thermal contact in the precooling stage (A $\rightarrow$ B) but which can give near perfect isolation in the adiabatic cooling stage (B $\rightarrow$ C). Mechanical systems give too much frictional heating, and use of low pressure helium gas as a heat exchange medium has serious drawbacks (e.g. it is difficult to remove for the second stage). The modern solution is to use a thermal link of high-conductivity metal (silver or copper) broken by a superconducting heat switch. This consists of a pure metal (aluminium or tin) which is a superconductor. When a (fairly small) magnetic field is applied to the switch, the superconductivity is destroyed and the metal is a good conductor of heat, as wanted in the precooling leg. However, when this field is removed and the switch becomes superconducting, it becomes a very poor conductor of heat – the superconducting electrons move in an ordered way, so that although they carry charge without resistance they have no entropy to carry! A superconducting heat switch can have a conductance ratio of 10^5 or higher at 10 mK.

Suppose we wish to cool liquid ^{3}He into the μK temperature region, a worthwhile task since it becomes a 'superfluid' below 1–2 mK. Clearly we can use the arrangement of Fig. 3.6 to cool some Cu spins. The major technical problem remaining is to achieve adequate thermal contact between the Cu spins and the sample of liquid ^{3}He. And this turns out to be the most severe problem of all. The trick is to cut down extraneous heat inputs (to the pW level) with antivibration measures and with careful electrical screening, and also to maximize the area of contact between the copper metal and the liquid with ultrafine sintered silver powder. In this way it is possible, at the time of writing, to achieve helium temperatures of below 90 μK.

What limits the final temperature? This is a question worth asking. The problem can be stated simply. If I take (3.7) seriously, then is it not possible to make the final temperature T_2 as small as I wish, simply by reducing B_2 to zero? In fact can the spins not be cooled to the absolute zero?

The answer is 'no, they cannot', one of many ways of stating the third law of thermodynamics. And the reason lies in the meaning of (3.7). That equation arose from the fact that the occupation numbers must be unchanged in an adiabatic process, i.e. that $2\mu B/k_B T$ must remain at a fixed value. The numerator $2\mu B$ is the energy level spacing ε, and it is this quantity which cannot reach zero. The two spin states can never have the same energy ('the ground state of a system can never be degenerate' might be another way of stating the third law). Instead there will always be some residual interactions left, even if these are only magnetic dipole–dipole forces between the N spins. As remarked earlier, although our treatment neglects interactions between particles some interactions are essential, otherwise thermal equilibrium could never be reached: hence 'weakly interacting'. We may turn off the externally applied magnetic field (B_0 say), but some residual interactions must remain. If we characterize these by an effective field (B_{int}), then the B entering equations such as (3.7) should be the appropriate sum of B_0 and B_{int}. Bearing in mind the random direction of B_{int}, the correct expression is

$$B = (B_0^2 + B_{int}^2)^{1/2} \tag{3.8}$$

The energy levels obtained from (3.8) are plotted in Fig. 3.7.

The upshot of all this on the lowest final temperature is:

1. The absolute zero is unattainable; the temperature reached in zero final applied field is $T_1(B_{int}/B_1)$, a non-zero number.
2. Another way of expressing this result is to observe that the spins will order in zero applied field at around a temperature $T_{int} = \mu B_{int}/k_B$. Referring back to Fig. 3.5, the entropy does not stay at $Nk_B \ln 2$ as T is lowered past this value, but falls to zero to achieve $S = 0$ at $T = 0$ (yet another statement of the third law).
3. Since the cooling method is useful only around $\mu B/k_B T \sim 1$, the lowest attainable practical temperature is around T_{int}. For CMN this is about 1 mK, a very low value for an electronic spin system; hence the importance of this particular salt. For Cu spins in the pure metal, T_{int} is less than 0.1 μK, one of the reasons for its choice.

3.1.4 Magnetization and thermometry

Before leaving the topic of the spin-$\frac{1}{2}$ solid we discuss its magnetic properties. These may be readily derived from the Boltzmann distribution, and they give a convenient method for measuring the low temperatures reached by adiabatic demagnetization.

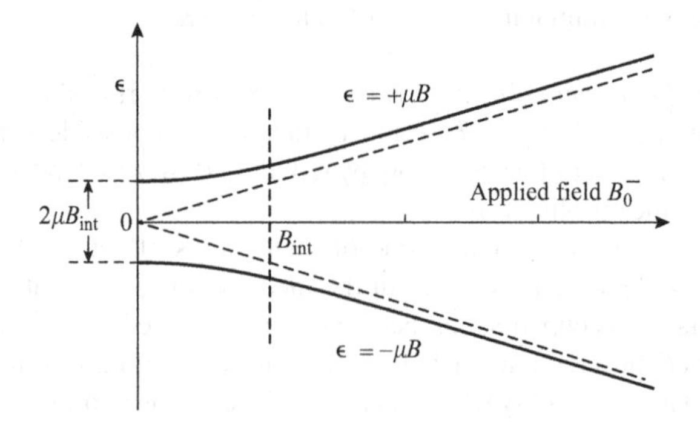

Fig. 3.7 The energies of a spin-$\frac{1}{2}$ solid as a function of applied magnetic field B_0.

Without labouring the point, it is no accident that the working substance of a refrigerator is also a possible candidate for thermometry – the refrigerant must have S a strong function of T, i.e. it must be thermally active ('doing something') in that temperature range!

Magnetization We characterize the magnetic properties by the magnetic moment M of the whole assembly of N spins. Interactions will be neglected, at any rate until we discuss ferromagnetism much later in the book, so that B may be taken as the externally applied field. In the lower state, each spin has a component μ aligned to the B field, whereas each spin in the upper state has a component $-\mu$. And the numbers of spins in the two states in thermal equilibrium at temperature T are given by the Boltzmann distribution (3.2). Hence we have

$$\begin{aligned} M &= n_0(\mu) + n_1(-\mu) \\ &= N\mu \frac{[\exp(\mu B/k_B T) - \exp(-\mu B/k_B T)]}{[\exp(\mu B/k_B T) + \exp(-\mu B/k_B T)]} \\ &= N\mu \tanh(\mu B/k_B T) \end{aligned} \tag{3.9}$$

This expression is plotted in Fig. 3.8, and one can note:

1. At high enough B, or low enough T, the magnetization saturates at the value $N\mu$, an obvious result since then all the N spins have aligned with the field.
2. For a solid with a higher spin (like the spin-$\frac{3}{2}$ Cu nuclei) the shape of the curve is superficially similar, although it is not a simple tanh function.
3. In the weakly paramagnetic limit, i.e. $\mu B/k_B T \ll 1$, at low fields or high temperatures, then the curve is essentially linear and we have:

$$M = N\mu(\mu B/k_B T) \tag{3.10}$$

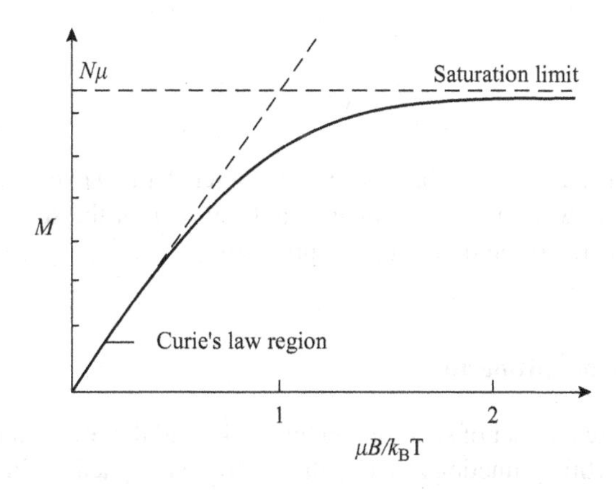

Fig. 3.8 The magnetization of a spin-$\frac{1}{2}$ solid as a (universal) function of $\mu B/k_B T$.

This result may readily be verified by expanding the exponentials in (3.9), bearing in mind that the exponents are small. Hence in this limit the magnetic susceptibility, essentially M/B, is proportional to $1/T$. This is *Curie's law*.

Therefore we have a direct $1/T$ thermometer. For an electronic paramagnet like CMN, the magnetization is usually measured directly from the total magnetization of the solid. The usual method is to measure the mutual inductance between two coils, wound over the CMN, giving a constant reading (which can be subtracted) and a component proportional to $1/T$. This may be used as a thermometer from 10 K or higher to around 1 mK, where interactions take over and spoil the simple story. Below about 20 mK, however, a nuclear system (often Pt nuclei) becomes useful. However, the value of μ is so small that the direct method is impractical – one would measure the effect of minute magnetic impurities with electronic spins. Therefore a resonance method is used (pulsed NMR) which singles out the particular energy level splitting of interest by means of radio frequency photons of the correct frequency. The strength of the NMR signal is directly proportional to $1/T$.

3.2 LOCALIZED HARMONIC OSCILLATORS

A second example which can be solved with little computational difficulty is that of an assembly of N localized harmonic oscillators. Suppose the oscillators are each free to move in one dimension only and that they each have the same classical frequency ν. They are therefore identical localized particles, and the results of Chapter 2 may be used to describe the equilibrium properties of the assembly at temperature T.

First we need the result from quantum mechanics for the states of one particle. For a simple harmonic oscillator there is an infinite number of possible states ($j = 0, 1, 2, 3, \ldots$) whose energies are given by

$$\varepsilon_j = \left(j + \frac{1}{2}\right) h\nu \qquad (3.11)$$

(Most books on quantum mechanics, e.g. Chapter 4 of Davies and Betts' book, *Quantum Mechanics*, in this series, include a discussion of the states of a harmonic oscillator, if you are not familiar with the problem.)

3.2.1 The thermal properties

There is an infinite number of states given by (3.11), and therefore an infinite number of terms in the partition function Z. Nevertheless the even spacing of the energy levels enables Z to be summed explicitly as a geometric progression. We have

$$\begin{aligned} Z &= \sum_j \exp\left[-\left(j+\frac{1}{2}\right)h\nu/k_\mathrm{B}T\right] && \text{from (2.24) and (3.11)} \\ &= \sum_j \exp\left[-\left(j+\frac{1}{2}\right)\theta/T\right] && \text{defining } \theta \text{ from } h\nu = k_\mathrm{B}\theta \\ &= \exp(-\theta/2T)\cdot\sum_j \exp(-j\theta/T) && \text{separating the common factor} \\ &= \exp(-\theta/2T)\cdot[1-\exp(-\theta/T)]^{-1} && \text{doing the sum (see below!)} \\ &= Z(0)\cdot Z(\mathrm{th}) && \text{say, compare (3.1)} \end{aligned} \tag{3.12}$$

The evaluation of the sum in the $Z(\mathrm{th})$ factor above is straightforward. If we write $y = \exp(-h\nu/k_\mathrm{B}T)$, then the required sum is $(1 + y + y^2 + y^3 + \cdots)$, which is readily summed to infinity to give $(1-y)^{-1}$. (Check it by multiplying both sides by $1-y$ if you are unsure). The separation in (3.12) into a zero-point term $Z(0)$ and a thermal term $Z(\mathrm{th})$ has a similar significance to our earlier factorization of (3.1).

Now that Z is evaluated in terms of T and the oscillators' scale temperature θ, there is no problem in working out expressions for the thermal quantities. For example, U is obtained by taking logarithms of (3.12) and differentiating as in (2.27). The result

$$U = \frac{1}{2}Nh\nu + \frac{Nh\nu}{\exp(\theta/T)-1} \tag{3.13}$$

is plotted in Fig. 3.9. The first term is the zero-point energy, derived from $Z(0)$, and the second term is the thermal energy arising from $Z(\mathrm{th})$. At high temperatures, as the figure demonstrates, the internal energy becomes

$$U = Nk_\mathrm{B}T \tag{3.14}$$

to a rather high degree of accuracy. (Expand the exponential in (3.13) to test this!) We return below to this simple but important expression.

The heat capacity is obtained by differentiating (3.13)

$$C = \frac{\mathrm{d}U}{\mathrm{d}T} = Nk_\mathrm{B}\frac{(\theta/T)^2\exp(\theta/T)}{[\exp(\theta/T)-1]^2} \tag{3.15}$$

This is plotted in Fig. 3.10, and as expected from (3.14) one may observe that $C = Nk_\mathrm{B}$ at high temperatures ($T > \theta$).

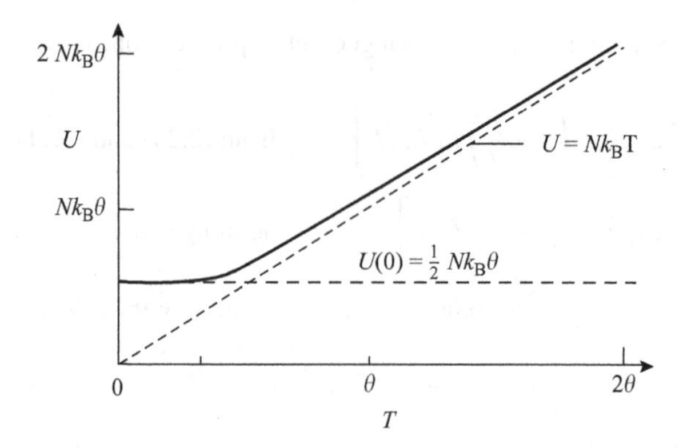

Fig. 3.9 The variation of internal energy with temperature for an assembly of harmonic oscillators. The characteristic temperature θ depends on the frequency of the oscillators (see text). Note the high and low temperature limits.

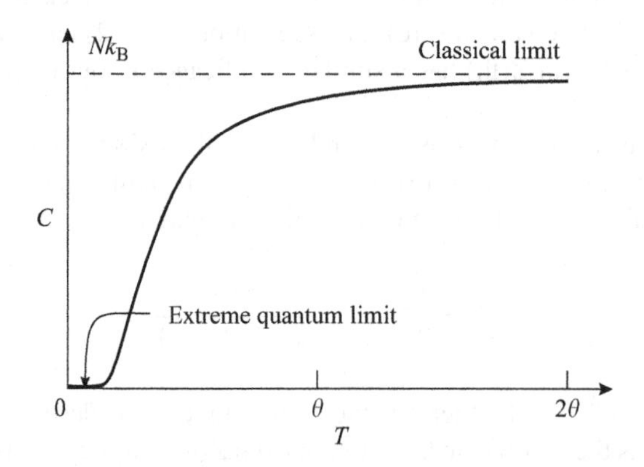

Fig. 3.10 The variation of heat capacity with temperature for an assembly of harmonic oscillators.

The entropy S is also readily derived, the easiest route again being to evaluate F, (2.28), and to differentiate F with respect to T. The result is

$$S = Nk_B \left[\ln \left(\frac{\exp(\theta/T)}{\exp(\theta/T) - 1} \right) + \frac{(\theta/T)}{(\exp(\theta/T) - 1)} \right] \tag{3.16}$$

The reader may wish to check that the low temperature limit is $S = 0$ (no surprise), and that at high temperatures the expression becomes $S = Nk_B \ln(T/\theta) = Nk_B \ln(k_B T/h\nu)$.

3.2.2 The extreme quantum and the classical limits

The assembly of localized oscillators forms a good example of the use of k_BT as the thermal energy scale.

Firstly consider the oscillators at an extremely low temperature, i.e. $T \ll \theta$ or $k_BT \ll h\nu$. The oscillators are frozen into the ground state, in that virtually none is thermally excited at any one time. We have $U = U(0)$, $C = 0$, $S = 0$. This situation should be thought of as the 'extreme quantum limit'. The discrete (i.e. quantum) nature of the states given by (3.11) is totally dominating the properties.

However, at the opposite extreme of temperature, i.e. $T \gg \theta$ or $k_BT \gg h\nu$, we reach a 'classical limit'. In this limit the simple expressions for U and C involve only k_BT and *not* $h\nu$. Planck's constant, i.e. the scale of the energy level splitting, is irrelevant now. Oscillators of any frequency have the same average energy k_BT. This dependence of U on k_BT alone is associated with the old classical 'law of equipartition of energy', which states that each so-called degree of freedom of a system contributes $\frac{1}{2}k_BT$ to the internal energy. In this case each oscillator displays two degrees of freedom (one for its kinetic energy and one for its potential energy), and the old law gives the result $U = Nk_BT$ for the N oscillators. But the question left unanswered by the classical law alone is: When is a degree of freedom excited and when is it frozen out? The answer is in the energy (or temperature) scales! We shall return to these ideas in our discussions of gases in Chapters 6 and 7.

Before leaving this topic, let us examine the entropy S. One can see from (3.16) that even in the classical limit, the expression for S involves h, Planck's constant. Entropy is fundamentally a quantum effect. From the classical region alone one can deduce that $S = S_0 + Nk_B \ln T$, but there is no way of finding the constant S_0 without integrating the thermal properties upwards from $T = 0$, i.e. through the quantum region. This problem lies at the heart of a number of the historical paradoxes and controversies of classical statistical mechanics in the pre-quantum age.

This quantum nature of S is well illustrated from a microscopic interpretation of (3.16). If one has an assembly of particles, each of which may be in G states with equal probability, then S is readily evaluated. ($G = 2$ is the spin-$\frac{1}{2}$ solid at high temperatures, for example, and we used $G = 4$ for Cu nuclei having spin $\frac{3}{2}$.) Since $\Omega = G^N$ for this situation, we may write $S = k_B \ln \Omega = Nk_B \ln G$. If we compare this expression with the high temperature limit of (3.16), we see that the two give the same answer if $G = k_BT/h\nu$, the ratio of the two energy scales. This is a pleasingly plausible result. It means that the entropy of the oscillators is the same at high temperatures as that of a hypothetical equal-occupation system in which G states are occupied. As k_BT rises, G continues to rise as higher levels (there are an infinite number in all!) come into play. Hence the $\ln T$ term. But the actual value of G depends on the ratio of k_BT to $h\nu$. It involves the quantum energy-level scale of the oscillators, so that G and S are inevitably quantum properties.

3.2.3 Applications

1. *Vibrations of solids* The assembly of identical harmonic oscillators was used by Einstein in the early days of quantum theory as a model for the thermal lattice vibrations of a simple atomic solid. He derived the result for the heat capacity (Fig. 3.10) as a universal function of (T/θ) with θ the so-called Einstein temperature characteristic of the solid. If we take a mole of solid, with N_A atoms, then, since each atom can vibrate in three dimensions, we should represent it by $N = 3N_A$ one-dimensional oscillators.

   The Einstein model has good and bad points. The successes are:

   1. It gives the correct high-temperature (classical) limit, namely $C = Nk_B = 3N_A k_B$ per mole. From the viewpoint of the previous section, this success arises simply from the correct evaluation of the number of degrees of freedom of the solid.
   2. It gives the correct extreme quantum limit, namely $C = 0$ at $T = 0$, a mystery in classical physics.
   3. It is indeed found experimentally that the heat capacity per mole is a universal function of (T/θ) for all simple solids, i.e. adjustment of a single parameter θ makes all results similar.

   However, the bad news is that the form of (3.15) is grossly incorrect in the intermediate temperature regime. In particular the experimental lattice heat capacity is not frozen out so rapidly as the theory predicts at low temperatures, but rather is proportional to T^3 in crystalline solids. The reason for the poor showing of the Einstein model is that by no stretch of imagination does a solid consist of localized oscillators which are weakly coupled. If one atom of the solid is moved, a disturbance is rapidly transmitted to the whole solid; and in fact the key to a more correct treatment is to model the vibrations as a gas of ultra-high frequency *sound* waves (phonons) in the whole solid, a topic to which we return in Chapter 9.
2. *Vibrations of gas molecules* The thermal vibrations of the molecules in a diatomic gas will be discussed in Chapter 7. In this case, rather unexpectedly, the Einstein model applies very well. Certainly the vibrations of one molecule are now weakly coupled from the vibrations of another molecule in a gas. However, our statistical treatment so far is appropriate to *localized* particles only, whereas gas molecules are certainly non-localized and in consequence indistinguishable. Not surprisingly, it is to a discussion of gases which we now turn.

3.3 SUMMARY

This chapter discusses the application of the Boltzmann distribution to derive the thermal properties of two types of substance, a spin-$\frac{1}{2}$ solid and an assembly of harmonic oscillators. We consider the somewhat idealized case in which the substance is modelled as an assembly of weakly-interacting localized particles, for which simple analytical solutions can be made.

1. For both substances, the one-particle energy states are well described by quantum mechanics.
2. A spin-$\frac{1}{2}$ particle has just two quantum energy levels, dependent on any applied magnetic field (Zeeman effect).
3. A harmonic oscillator has an infinite number of equally-spaced energy states.
4. The partition function $Z(T)$ is readily found in both cases. It can be expressed as a zero-point factor (which depends on the energy zero but not on temperature) multiplied by a thermal factor (which is independent of the energy zero but does depend on temperature).
5. The thermal properties are dependent on the ratio of two energy scales. One derives from the one-particle energy structure, the other is $k_B T$, the 'thermal energy scale'.
6. The entropy rises from zero at low temperature to approximately $Nk_B \ln G$ at high temperature, where G is the number of one-particle states accessed at the high temperature. For the spin-$\frac{1}{2}$ solid $G = 2$, independent of temperature; whereas for the oscillators G is proportional to T.
7. Cooling by adiabatic demagnetisation of a paramagnetic material is readily understood from the properties of a spin-$\frac{1}{2}$ solid.
8. Harmonic oscillators provide a model (the Einstein model) for understanding some aspects of the thermal properties of solids, even though the atoms of a solid are hardly 'weakly-interacting'. We shall return to this question later in Chapter 9.

4

Gases: the density of states

In the last two chapters we have applied the statistical method as outlined in section 1.5 to an assembly of distinguishable (localized) particles. We now embark upon the application of the same method to gases. This involves two new ideas.

The first concerns the one-particle states (step I of section 1.5). A gas particle is confined to a large macroscopic box of volume V, whereas a localized particle is confined essentially to an atomic cell of volume (V/N). As a direct result, the energy levels of a gas particle will be extremely close together, with astronomical numbers of them being occupied at any reasonable temperature. This is in marked contrast with the localized case, in which the energy levels are comparatively far apart and the occupation of only a few need to be considered. Actually it turns out that one can make a virtue of the necessity to consider very large numbers of states. This topic is the subject of the present chapter.

The second idea is how to deal with the indistinguishability of gas particles. This is vital to the counting of microstates (step III of section 1.5). The ideas are essentially quantum mechanical, explaining from our modern viewpoint why a correct microscopic treatment of gases was a matter of such controversy and difficulty in Boltzmann's day. The counting of microstates and the derivation of the thermal equilibrium distribution for a gas will be discussed in the following chapter.

4.1 FITTING WAVES INTO BOXES

We are going to be able to discuss many types of gaseous assemblies in the next few chapters, from hydrogen gas to helium liquid, from conduction electrons to black-body radiation. However, there is a welcome economy about the physics of such diverse systems. In each case, the state of a gas particle can be discussed in terms of a wavefunction. And the basic properties of such wavefunctions are dominated by pure geometry only, as we will now explain.

Consider a particle in one of its possible states described by a wavefunction $\psi(x, y, z)$. The time-dependence of the wavefunction need not be explicitly included here, since in thermal physics we only need the stationary states (i.e. the normal modes or the eigenstates) of the particle. What do we know about such states when the particle is confined within an otherwise empty box of volume V? For convenience, let us assume that the box is a cube of side a (so $V = a^3$). Furthermore we assume that 'confined' means that it must satisfy the boundary condition $\psi = 0$ over the surface of the box. Now we come to the 'economy' referred to above: we know that the wavefunction ψ is a simple sinusoidal oscillation inside the box. This simple statement is a correct summary of the solution of Schrödinger's equation for a particle of mass m, or of the wave equation for a sound or electromagnetic wave. Therefore the only way to achieve a possible wavefunction is to ensure that a precisely integral number of half-waves fit into the box in all three principal directions. Hence the title of this section! The problem is the three-dimensional analogue of the one-dimensional standing-wave normal modes of the vibrations of a string.

To be specific we choose the origin of co-ordinates to be at one corner of the box. The wavefunction is then given by a standing wave of the form

$$\psi \sim \sin(n_1 \pi x/a) \cdot \sin(n_2 \pi y/a) \cdot \sin(n_3 \pi z/a) \tag{4.1}$$

where the positive integers $n_{1,2,3}$ $(= 1, 2, 3, 4 \ldots)$ simply give the number of half-waves fitting into the cubical box (side a) in the x, y and z directions respectively. (The use of sine functions guarantees the vanishing of ψ on the three faces of the cube containing the origin; the integral values of the three ns ensures the vanishing of ψ on the other three faces.) Incidentally, in terms of our earlier notation the state specification by the three numbers (n_1, n_2, n_3) is entirely equivalent to the previous state label 'j'.

It is useful to write (4.1) in terms of the components of a 'wave vector' k as

$$\psi \sim \sin(k_x x) \cdot \sin(k_y y) \cdot \sin(k_z z) \tag{4.2}$$

where

$$\boldsymbol{k} = (k_x, k_y, k_z) = (\pi/a)(n_1, n_2, n_3)$$

The possible states for the wave are specified by giving the three integers $n_{1,2,3}$, i.e. by specifying a particular point $\boldsymbol{k}$ in 'k-space'. This is a valuable geometrical idea. What it means is that all the possible 'k-states' can be represented (on a single picture) by a cubic array of points in the positive octant of k-space, the spacing being (π/a).

Now because of the macroscopic size a of the box, this spacing is very small in all realistic cases. Indeed we shall find later that in most gases the states are so numerous that our N particles are spread over very many more than N states – in fact most occupation numbers are 0. Under these conditions it is clear that we shall not wish to consider the states individually. That would be to specify the state of the assembly in far too much detail. Rather we shall choose to *group* the states, and to specify only the mean properties of each group. And the use of k-space gives an immediate way of doing the grouping (or graining as it is often called).

For instance let us suppose we wish to count the number of states in a group whose $\boldsymbol{k}$ lies with x-component between k_x and $k_x + \delta k_x$, with y-component between

k_y and $k_y + \delta k_y$ and with z-component between k_z and $k_z + \delta k_z$. Call this number $g(k_x, k_y, k_z)\delta k_x \delta k_y \delta k_z$. Since the states are evenly spread in k-space with the spacing (π/a), the required number is

$$g(k_x, k_y, k_z)\delta k_x \delta k_y \delta k_z = (a/\pi)^3 \delta k_x \delta k_y \delta k_z \tag{4.3}$$

The function g so defined is a 'density of states' in $\boldsymbol{k}$. Its constant value simply sets out that the states are uniformly spread in k-space. However, this is a significant and useful idea. The number of states in any group can be evaluated from the volume in k-space occupied by the group, together with (4.3).

For example, to discuss the equilibrium properties of gases we shall make extensive use of another density of states function $g(k)$, which relates to the (scalar) magnitude of $\boldsymbol{k}$ only, irrespective of its direction. Its definition is that $g(k)\delta k$ is the number of k-states with values of (scalar) k between k and $k + \delta k$. Its form may be derived from (4.3) from an integration over angles. Since the density of states in $\boldsymbol{k}$ is constant, the answer is simply

$$\begin{aligned} g(k)\delta k &= (a/\pi)^3 \cdot \text{appropriate volume in } k\text{-space} \\ &= (a/\pi)^3 \cdot (4\pi k^2 \delta k/8) \\ &= V/(2\pi)^3 \cdot 4\pi k^2 \delta k \end{aligned} \tag{4.4}$$

In this derivation the factor $4\pi k^2 \delta k$ arises as the volume of a spherical shell of radius k and thickness δk. The $\frac{1}{8}$ factor comes since we only require the $\frac{1}{8}$ of the shell for which k_x, k_y and k_z are all positive (i.e. the positive octant).

Equation (4.4) is the most important result of this section. Before discussing how it is used in statistical physics, we may note several points.

1. The dependence on the box volume V in (4.4) is always correct. Our outline proof related to a cube of side a. However, the result is true for a box of any shape. It is easy to verify that it works for a cuboid – try it! – but it is not at all easy to do the mathematics for an arbitrary shape of box. However, a physicist knows that only the volume (and not the shape) of a container of gas is found to influence its bulk thermodynamic properties, so perhaps one should not be surprised at this result.
2. Rather more interestingly, the result (4.4) remains valid if we define differently what we mean by a box. Above we adopted standing wave-boundary conditions, i.e. $\psi = 0$ at the box edge. However, this boundary condition is not always appropriate, just because it gives standing waves as its normal modes. In a realistic gas there is some scattering between particles, usually strong enough that the particles are scattered many times in a passage across the box. Therefore it is not always helpful to picture the particles 'rattling about' in a stationary standing-wave state. This is particularly true in discussing transport properties or flow properties of the gas. Rather one wishes a definition of a box which gives travelling waves for the normal modes, and this is achieved mathematically by adopting'periodic boundary conditions' for a cuboidal box.

In this method, the periodic condition is that both ψ and its gradient should match up on opposite faces of the cube. (One way of picturing this is that if the whole of space were filled with identical boxes, the wavefunction would be smooth and would be exactly replicated in each box.) The normal modes are then travelling waves of the form

$$\psi \sim \exp(ik_x x + ik_y y + ik_z z)$$

or

$$\psi \sim \exp(i\boldsymbol{k}\cdot\boldsymbol{r})$$

The values of $\boldsymbol{k}$ differ from the earlier ones in two respects, and indeed the microscopic picture is quite different in detail. The first difference is that the spacing of the k-values is doubled, since now the boundary condition requires fitting an integral number of full (and not half) waves into the box. Hence the spacing between allowed k-values becomes $(2\pi/a)$, and (4.3) is no longer valid – another factor of $(\frac{1}{8})$ is needed on the right-hand side. The second difference is that the restriction on $\boldsymbol{k}$ to the positive octant is lifted. A negative k-value gives a different travelling wave state from positive k; they represent states with the same wavelength but travelling in opposite directions. This difference means that the integration between (4.3) and (4.4) should cover all eight octants of k-space. Therefore the two differences compensate for each other in working out $g(k)\delta k$ and the result (4.4) survives unchanged.

3. The next comment concerns graining. The states are not evenly spread in k-space on the finest level – they are discrete quantum states. Therefore we cannot in principle let the range δk in (4.4) tend to zero. The group of states we are considering must always be a large one for the concept of the (average) density of states to make sense. Nevertheless, in every practical case except one (the Bose–Einstein condensation to be discussed in Chapter 9) we shall find that the differences between adjacent groups can still be so minutely small, that calculus can be used, i.e. we can for computation replace the finite range δk by the infinitesimal $\mathrm{d}k$.
4. The final comment is about dimensionality. Equation (4.4) is a three-dimensional result, based on the volume of a spherical shell of radius k. Entirely analogous results can be derived for particles constrained within one- or two-dimensional 'boxes', a topic of much importance to modern nanoscience. For example consider the states of a particle constrained in a two-dimensional sheet. The confinement happens because the width of the box normal to the sheet is made so small that the wavefunction in that direction is fixed to be lowest standing wave state (one half wave across the sheet), with the next state (two half waves) out of thermal energy range. The wavefunctions within the sheet can again be treated as extended travelling waves. The particles in a sheet of area A have a density of states in k of

$$g(k)\delta k = A/(2\pi)^2 \cdot 2\pi k\delta k$$

The corresponding one-dimensional result for a line (a 'quantum wire') of length L is rather similar to (4.3) since no integration is needed

$$g(k)\delta k = L/(2\pi)\delta k$$

The reader might care to derive these results.

4.2 OTHER INFORMATION FOR STATISTICAL PHYSICS

In the previous section we have described the geometry of fitting waves into boxes, a common feature for all types of gaseous particle. However, before we know enough about the states to proceed with statistical physics, there are two other ideas which need to be stirred in. Although the final answers will be different for different gases, the same two questions need to be asked in each particular case.

4.2.1 Quantum states are k-states plus

The first question relates to what constitutes a quantum state for a particle. The general answer is that the full state specification (labelled simply by j in earlier chapters) must include a complete statement of what can in principle be specified about the particle. The k-state of the particle (as given by the three numbers $n_{1,2,3}$ for our cubical box) is a full specification only of the translational motion of the centre of mass of the particle, as characterized by $\psi(x, y, z)$.

But usually there are other identifiable things one can say about the particle, besides its centre of mass motion. Firstly, one may need to consider internal motion of the particle, arising from vibration or rotation of a molecular particle for example. We shall return to this type of 'internal degree of freedom' in Chapter 7. Secondly, even for a simple particle, one must specify the spin of the particle, or the polarization of the corresponding wave, essentially the same idea. For example an electron with a particular spatial wavefunction and its spin-up is in a *different* quantum state from an electron with the same spatial wavefunction but its spin-down (hence the periodic table, for instance!). Similarly for an electro-magnetic wave one needs to know not only its wave vector $\boldsymbol{k}$ but also its polarization (e.g. left- or right-handed).

The idea of spin may easily be included in (4.4). We redefine $g(k)\delta k$ to be the number of quantum states (and not merely k-states) with magnitude of k between k and $k + \delta k$. The equation then becomes

$$g(k)\delta k = V/(2\pi)^3 \cdot 4\pi k^2 \delta k \cdot G \tag{4.5}$$

where the new factor G is a polarization or spin factor, usually 1 or 2 or another small integer, dependent on the type of substance under consideration. In future all density of states functions like $g(k)$ will refer to quantum states and will include this spin factor.

4.2.2 The dispersion relation

The second vital idea is that to make any progress with statistical physics one must not only be able to count up the quantum states. One must also know their energies, ε_j in earlier notation. The point is simply that all equilibrium thermal properties are governed only by energy; the two constraints on an allowed distribution are those of energy and number.

In order to determine the energy ε_j, there are two considerations. The first of these is the 'plus' of the previous section. If there are internal spatial degrees of freedom, these will affect the energy (see Chapter 7). In addition, sometimes the energy is spin-dependent, for example in the case of an electron in a magnetic field, and then this must be included and it will have important effects. But the second and more general consideration is that in every case the energy depends on k. We use $\varepsilon(k)$, or simply ε, in the rest of this chapter to represent the energy contribution from the translational motion, i.e. from the k-state. The $\varepsilon - k$ relation is often referred to as the dispersion relation. It contains precisely the same physical content as the $\omega - k$ dispersion relation in wave theory (since $\varepsilon = \hbar\omega$) and the $\varepsilon - p$ energy-momentum relation in particle mechanics (since $p = \hbar k$).

It is usually a convenience in statistical physics to combine the geometry of k-space (equation (4.5)) with the $\varepsilon - k$ relation to give a 'density of states in energy', often called simply the density of states. This quantity is defined so that $g(\varepsilon)\delta\varepsilon$ is the number of states with energies between ε and $\varepsilon + \delta\varepsilon$. It is derived from (4.5) by using the dispersion relation to transform both the k^2 factor and the range δk. This point is worth stressing. It is always worth writing a density of states function with its range explicitly shown, as for example on both sides of (4.3), (4.4) and (4.5). Although the same symbol g is used for $g(k)$ and for $g(\varepsilon)$, these two functions have different dimensions and units. It is the functions multiplied by their ranges, i.e. $g(k)\delta k$ and $g(\varepsilon)\delta\varepsilon$, which have the same dimensions – they are pure numbers. An example of this transformation is given in the next section.

4.3 AN EXAMPLE – HELIUM GAS

As a specific example of the ideas of this chapter, let us consider helium (^{4}He) gas contained in a 10-cm cube at a temperature of 300 K.

The states are given by the solution of the time-independent Schrödinger equation for a particle of mass M in a field-free region

$$(-\hbar^2/2M)\nabla^2\psi = \varepsilon\psi \tag{4.6}$$

with boundary conditions as discussed in section 4.1. The solutions are precisely those of (4.1) and (4.2). Substitution of these solutions back into (4.6) gives for the energies of the particle

$$\varepsilon = \hbar^2 k^2/2M \tag{4.7}$$

i.e.

$$\varepsilon = (h^2/8Ma^2)(n_1^2 + n_2^2 + n_3^2) \tag{4.8}$$

The energy $(h^2/8Ma^2) = \Delta\varepsilon_j$, say, gives a scale for the energy level spacing for a helium atom in the gas. It works out at 8×10^{-40} J, a very small energy compared with $k_B T$ which equals 4×10^{-21} J at 300 K. Clearly $k_B T \gg \Delta\varepsilon_j$, and from what we know already about energy scales this implies that a very large number of states will be energetically available for occupation in the thermal equilibrium state. Hence the whole approach of this chapter.

Finally we derive the density of states. Helium is a monatomic gas, so (4.5) gives the density of states in k. In this instance, since ^{4}He has zero spin, the factor $G = 1$. Hence $g(k)$ is known. The appropriate dispersion relation (from the Schrödinger equation) is (4.7), so the density of states in ε can be determined, treating k and therefore ε as continuous variables in view of the large numbers of states involved.

The calculation goes as follows. Starting from (4.7), $\varepsilon = \hbar^2 k^2/2M$, differentiation gives

$$\delta\varepsilon \quad = \hbar^2 k\delta k/M$$

and inversion gives

$$k = (2M\varepsilon/\hbar)^{1/2}$$

Equation (4.5) with $G = 1$ is

$$g(k)\delta k = V/(2\pi)^3 \cdot 4\pi k^2 \delta k$$

Substituting for k and for $k\delta k$ we obtain

$$\begin{aligned} g(\varepsilon)\delta\varepsilon &= V/(2\pi)^3 \cdot 4\pi(2M\varepsilon/\hbar^2)^{1/2}(M\delta\varepsilon/\hbar^2) \\ &= V2\pi(2M/h^2)^{3/2}\varepsilon^{1/2}\delta\varepsilon \end{aligned} \tag{4.9}$$

This derivation nicely demonstrates the transformation from k to ε and the final result (4.9) will be a useful one in our later discussions of gases.

4.4 SUMMARY

This chapter lays the groundwork for later discussion of gases, by discussing the one-particle states.

1. Since a gas particle is free to roam over a whole macroscopic box, its possible states are very closely spaced.
2. The geometrical ideas of 'fitting waves into boxes' apply to any gaseous system.

3. States are uniformly distributed in k-space, that is the density of states in k-space is constant, depending only on the box volume V.
4. Therefore a grouping of states, needed to define a sensible distribution, is conveniently related to volume in k-space.
5. Similar ideas are used to describe gases in situations of restricted dimensions (quantum sheets or wires).
6. Quantum states are specified by their k-space properties, but also can have additional specifications because of spin or polarisation, introducing a further factor (G in (4.5)).
7. The density of states in energy plays an important role in thermal properties, and is worked out from the k-space ideas together with the appropriate dispersion (energy-momentum) relation .

5

Gases: the distributions

In this chapter the statistical method outlined in section 1.5 is used to derive the thermal equilibrium distribution for a gas. The results will be applied to a wide variety of physical situations in the next four chapters.

The method follows the usual four steps. Step I concerns the one-particle states. These were discussed in the previous chapter, and in fact no further specific discussion is called for until we come to particular applications. The one vital feature to recall here is that the states are very numerous. Hence we shall discuss gases in terms of a grouped distribution as explained in section 5.1. This discussion of possible distributions is step II of the argument. Step III involves counting microstates, i.e. quantum states for the N-particle assembly. In section 5.2 we briefly review the quantum mechanics of systems containing more than one identical particles. The counting of microstates is then outlined in section 5.3 and finally (step IV) we derive the thermal (most probable) distribution in section 5.4.

5.1 DISTRIBUTION IN GROUPS

As discussed in the last chapter, the number of relevant one-particle states in a gas is enormous, often very much greater than the number N of gas particles. Under such conditions, the definition of a distribution in states contains far too much detail for comfort. Instead we use effectively the distribution in levels, $\{n_i\}$ as defined in section 1.4.

The point is that it does no violence to the physics to *group* the states. In the form of the distribution used for computation the ith group is taken to contain g_i states of average energy ε_i. The only difference from the true specification of the states is that in the grouped distribution all g_i states are considered to have *the same* energy ε_i, rather than their correct individual values. This grouped distribution is illustrated in Fig. 5.1. Let us note several points before proceeding.

1. The grouping is one of convenience, not of necessity. We are choosing to use large numbers of states g_i taken together and also large numbers of particles n_i occupying these states. This will enable us below to use the mathematics of large numbers (e.g. Stirling's approximation) in discussing n_i and g_i in order to work

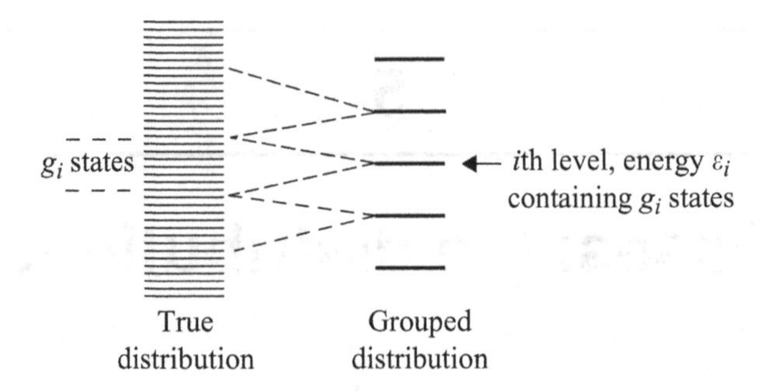

Fig. 5.1 Distribution in groups. The true distribution, containing very many states indeed, can be replaced by the grouped distribution for the purpose of calculation.

out the distribution. But as physicists we can rest assured that more sophisticated approaches give the same mathematical results even if the groups are not assumed large (or if the grouping is not carried out at all); however, why not accept a short-cut if one is offered?

2. The grouping used in the distribution is arbitrary, but not capricious! The rule is that the number of states remains the same in both sides of Fig. 5.1. Or, more precisely, the density of states (over a sufficiently coarse energy graining) remains unchanged. If we choose the energy levels ε_i, then the full microscopic details of the true physical problem determine the values g_i.
3. A test for whether the grouping makes physical sense is to consider the average number of particles per state, often called the *distribution function* or colloquially the *filling factor*. This is defined as

$$f_i = n_i/g_i$$

Effectively, this filling factor f_i tells us the fractional occupation of a state of energy ε_i. In other words, within the density of states approximation, it can be thought of purely as a function $f(\varepsilon)$ of the energy ε. The test just referred to is a simple one. If the grouping has not violated the original problem, we expect that f_i will be a well-defined function of ε_i whose value does *not* depend on the details of the grouping. In particular, it should not depend on the value of g_i.
4. Finally, if we return to the example of helium gas (section 4.3), we can see that this approach makes some numerical sense. We noted previously that the energy level spacing between the true states ($\Delta\varepsilon_j$) was about 10^{-19} times $k_B T$. Therefore, it is quite possible to use groups of, say, 10^{10} states each (so that $g_i \gg 1$ in anybody's language), whilst maintaining an energy level spacing $\Delta\varepsilon_i$ which is still minute (about 10^{-9} to 1) compared to the thermal energy scale $k_B T$.

5.2 IDENTICAL PARTICLES – FERMIONS AND BOSONS

The next step (step III) in the statistical method is to count the number of microstates corresponding to a particular distribution. A microstate of course relates to a quantum state of all N particles, and here must come, therefore, the vital recognition that identical gas particles are fundamentally indistinguishable from each other. They are, therefore, effectively competing for the same one-particle states. We can never define *which* particle is in a particular state, the microstate is completely specified merely by stating *how many* particles are in each state. This counting problem will be tackled in the next section, but first there is another important question to be addressed. In quantum mechanics there are two different types of identical particle, which obey different rules as to how many particles can be allowed in each state. The particle types are called fermions and bosons.

Consider a system of just two identical indistinguishable particles. It is described in quantum mechanics by a two-particle wavefunction $\psi(1, 2)$, in which the label 1 is used to represent all the co-ordinates (i.e. space and spin co-ordinates) of particle 1 and the label 2 represents the co-ordinates of the other particle. For $\psi(1, 2)$ to be a valid wavefunction, two conditions must be satisfied. One is obviously that it should be a valid solution of Schrödinger's equation. But the second condition is that it should satisfy the correct symmetry requirement for interchange of the two labels. The symmetry required, that of 'interchange parity', can be outlined as follows.

If the labels of the two particles are interchanged, then any physical observable cannot change (since the particles are identical). This implies that

$$\psi(1,2) = \exp(i\delta)\psi(2,1)$$

since a physical observable always involves the product $\psi^*\psi$. The wavefunctions are related by the phase factor $\exp(i\delta)$. But if we make two interchanges, then we come to a mathematical identity, i.e.

$$\psi(1,2) = \exp(i\delta)\,\psi(2,1) = \exp(2i\delta)\,\psi(1,2)$$

Hence $\exp(2i\delta) = 1$, and therefore the interchange factor $\exp(i\delta)$ must equal $+1$ or -1, one of the two square roots of $+1$.

Particles which interchange co-ordinates with the $+1$ factor are called 'bosons', and those with the -1 factor are 'fermions'. We shall use the symbol S to describe the symmetric wavefunction of bosons, and A for the antisymmetric function of fermions. It can be shown (not here!), or it can be taken as an experimental result, that all particles with integral spin (0, 1, 2, etc.) are bosons; whereas odd-half integral spin particles (spin $\frac{1}{2}, \frac{3}{2}$, etc.) are fermions. Therefore, for example, electrons and ^{3}He (both having spin $\frac{1}{2}$) are fermions; photons (spin 1) and ^{4}He (spin 0) are bosons.

Next, let us consider the case where the particles are weakly interacting, the relevant case for our statistical physics. This means that a solution to Schrödinger's equation for the two particles can be written as the product of two one-particle wavefunctions, i.e.

$$\psi(1,2) = \psi_a(1) \cdot \psi_b(2) \quad (5.1)$$

where the labels a and b are two values of the state label j for the one-particle states. However, as a wavefunction (5.1) suffers from the defect that it has no interchange parity, it is neither A nor S. Nevertheless it can be combined with its interchanged partner (an alternative solution to the Schrödinger equation with the same properties) to yield two wavefunctions, one S and one A. These are

$$\psi_S(1,2) = (1/\sqrt{2})[\psi_a(1)\psi_b(2) + \psi_a(2)\psi_b(1)] \quad (5.2)$$

$$\psi_A(1,2) = (1/\sqrt{2})[\psi_a(1)\psi_b(2) - \psi_a(2)\psi_b(1)] \quad (5.3)$$

The reader may readily check that these functions possess the correct parity, namely $\psi_S(1,2) = +\psi_S(2,1)$ and $\psi_A(1,2) = -\psi_A(2,1)$. The $(1/\sqrt{2})$ factors are for normalization purposes only.

Equations (5.2) and (5.3) bring out a fundamental difference between bosons and fermions. Consider the situation in which the labels a and b are identical, i.e. in which the two particles are both competing for the same state. For bosons, for which (5.2) applies, there is no problem. (In fact (5.2) gives an enhanced wavefunction compared to (5.1).) On the other hand fermions are quite different. The wavefunction given by (5.3) vanishes identically when we set $a = b$. This is the *Pauli exclusion principle*, which recognizes that 'no two identical fermions can occupy the same state'. I have heard bosons referred to as 'friendly particles' and fermions as 'unfriendly'. Although bosons enjoy multiple occupancy of states, in a fermion society all states are either unoccupied or singly occupied!

Finally we may readily generalize all the above results to an assembly containing N rather than merely two identical particles. The interchange argument is still valid between any pair of particles. Actually this makes it obvious that the choice of $+1$ or -1 must be a generic choice. When the first interchange considered is assigned a parity, an inconsistency will arise if all other interchanges are not given the same parity. All electrons are fermions, and all ^{4}He atoms are bosons.

For our assembly of N weakly interacting particles, the generalizations of (5.1), (5.2) and (5.3) are obvious but a little tedious to write down. Equation (5.1) becomes an extended product of N one-particle terms. For bosons, the expression for ψ_S is similar to (5.2), except that it contains $N!$ terms (every permutation of the particle co-ordinate labels) all combined with a + sign. The normalization factor becomes $(1/\sqrt{N!})$. Similarly for fermions, the generalization of (5.3) is the one antisymmetric arrangement of the $N!$ terms; this will have systematically alternating signs, and may be written neatly as a determinant in which rows give the state label and columns the particle co-ordinate. But the vital feature is that the exclusion principle as stated above still operates – if there are two particles in the same state ψ_A is identically zero.

5.3 COUNTING MICROSTATES FOR GASES

Now we are equipped to attack step III of the statistical method, that of counting up the number of microstates consistent with a given valid distribution. Since we postulate that all such microstates are equally probable, this step is the essential prerequisite to finding the most probable distribution, i.e. that which best describes thermal equilibrium, but that final step (step IV) follows in the next section.

The counting problem may be set up as follows. The indistinguishable nature of the particles is accounted for by the inclusion of all permutations of the particle co-ordinates in the generalizations of (5.2) and (5.3). The microstate (effectively the appropriate wavefunction either ψ_S or ψ_A) can therefore be labelled just by the one-particle state labels a, b, etc. As pointed out earlier, this is simply a recognition that we cannot know which particles are in which states; even in principle we can only know which states are occupied. Therefore the counting of microstates is to specify the occupation of each one-particle state.

The distribution whose number of microstates is to be counted is the grouped distribution defined in section 5.1. The required number $t(\{n_i\})$ can be written as the product of contributions from each group of states, but the form of these factors will differ for bosons and fermions. The treatments of fermions (with the exclusion principle) and of bosons (without) are quite different in principle, which goes two-thirds of the way to explaining why this section has three sub-sections. The reason for the third will emerge below.

5.3.1 Fermions

As we have seen in section 5.2, the exclusion principle operates for fermions. Therefore, the one-particle states can only have occupation numbers of 0 or 1. Incidentally, this implies that in the ith group, the number of particles n_i cannot exceed the number of states g_i.

The counting is now straightforward. The group of g_i states is divisible into two subgroups: n_i of the states are to contain one particle, and therefore the other $(g_i - n_i)$ must be unoccupied. This is the 'binomial theorem' counting problem of Appendix A, problem 2. The number of different ways the states can be so divided is

$$\frac{g_i!}{n_i!(g_i - n_i)!}$$

This is the contribution to the number of microstates from the ith group. The total number of microstates corresponding to an allowable distribution $\{n_i\}$ is therefore given by

$$t_{\text{FD}}(\{n_i\}) = \prod_i \frac{g_i!}{n_i!(g_i - n_i)!} \tag{5.4}$$

The symbol FD is short for Fermi–Dirac, since these two physicists were responsible for the first discussion of the statistics of what we now call fermions.

5.3.2 Bosons

The statistics for bosons is called Bose–Einstein (BE) statistics, also named after its inventors. Counting the boson microstates is a little more tricky than for fermions since now any number of particles are allowed to occupy any one-particle state.

A direct, if slightly abstract, way of calculating the contribution from the ith group of the distribution is as follows. In the group there are g_i states containing n_i identical particles with no restrictions on occupation numbers. A typical microstate can be represented as in Fig. 5.2 by $(g_i - 1)$ lines and n_i crosses. The lines represent divisions between the g_i states, and the crosses represent an occupying particle. New microstates representing the same distribution, i.e. the same value of n_i for the group, can be obtained by shuffling the lines and crosses on the picture. In fact the contribution to the number of microstates is precisely a further simple binomial problem: how many ways can the $(n_i + g_i - 1)$ symbols be arranged into n_i crosses and $(g_i - 1)$ lines? The answer is the binomial coefficient (see again Appendix A)

$$\frac{(n_i + g_i - 1)!}{n_i!(g_i - 1)!}$$

This is the correct answer to the problem. But in fact, bearing in mind that in statistical physics we are always dealing with large numbers g_i of states, the -1 addition is negligible. It is adequate to use g_i rather than $(g_i - 1)$, and we shall do this forthwith. (Actually, if the -1 is retained here, then in the next section the mathematics would give a distribution n_i proportional to $(g_i - 1)$ rather than to g_i violating the grouping requirement of section 5.1. The -1 would have then to be omitted for consistency, so we omit it from the outset!)

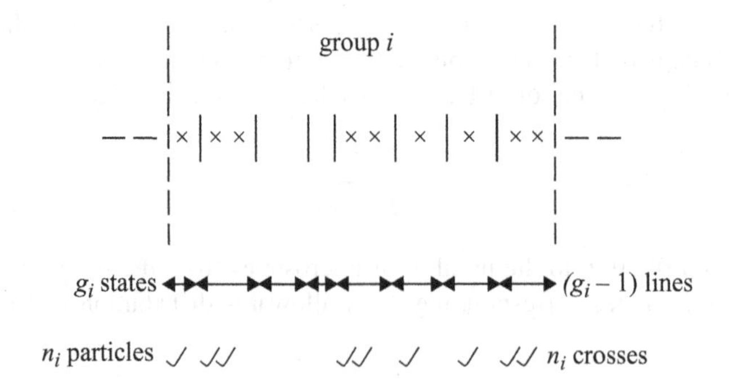

Fig. 5.2 Counting bosons. In each group the g_i states are represented by $(g_i - 1)$ lines and the n_i particles by n_i crosses. In the illustration a micro-state is represented for the case $g_i = 8, n_i = 9$.

Hence our (slightly approximate) final result for the number of microstates to an allowed distribution for a boson gas is

$$t_{\mathrm{BE}}(\{n_i\}) = \prod_i \frac{(n_i + g_i)!}{n_i! g_i!} \tag{5.5}$$

where again the product ranges over all groups i.

5.3.3 Dilute gases

For gases, there are two, and only two, possibilities. The gaseous particles are either bosons (+1, symmetric) or fermions (−1, antisymmetric). There is no half-way house. However, it is profitable to consider a third form of statistics quite explicitly, namely that of a dilute gas.

The word 'dilute' here has a specific meaning. It is to suppose that for all groups i, the states are very sparsely occupied, i.e.

$$n_i \ll g_i \text{ for all } i$$

We shall see later that this condition holds for real gases in the limit of low density and high temperature, and that it corresponds to a classical limit to either the FD or the BE form of quantum statistics.

In this dilute limit, we can readily see that both forms of statistics give almost the same answer for $t(\{n_i\})$. It is a result we should expect, since when $n_i \ll g_i$ even for bosons we would anticipate that almost all the states are unoccupied, and just a few have a single occupation. The existence or otherwise of the exclusion principle is irrelevant if the system is not even tempted towards multiple occupation!

To consider the fermion case first, each factor in (5.4) can be written as

$$[g_i(g_i - 1)(g_i - 2) \ldots (g_i - n_i + 1)]/n_i!$$

where both numerator and denominator have been divided by $(g_i - n_i)!$. Note that the numerator of this expression is the product of n_i factors, all of which in our present limit are almost equal to (but a bit less than) g_i. Therefore, we have

$$t_{\mathrm{FD}} \approx \prod_i \frac{g_i^{n_i}}{n_i!}$$

In the boson case, (5.5) can be treated similarly in the dilute limit. Dividing top and bottom by $g_i!$ one obtains for the typical term

$$[(g_i + 1)(g_i + 2) \ldots (g_i + n_i)]/n_i!$$

Again the numerator is the product of n_i factors each approximately equal to (but a little larger than) g_i. Hence for the dilute boson gas

$$t_{\mathrm{BE}} \approx \prod_i \frac{g_i^{n_i}}{n_i!}$$

Therefore in the dilute limit t_{FD} and t_{BE} tend to the same value, one from below and one from above. This 'classical' limit is called the Maxwell–Boltzmann (MB) case, discussed by these famous scientists before quantum mechanics was conceived. And we write the answer to our third counting problem as

$$t_{\mathrm{MB}} = \prod_i \frac{g_i^{n_i}}{n_i!} \tag{5.6}$$

5.4 THE THREE DISTRIBUTIONS

It now remains to derive the three equilibrium distributions for the three counting methods of section 5.3. The techniques are precisely those set up in section 1.5 and already worked through for localized particles in Chapter 2. The aim is to find the most probable distribution consistent with the macrostate by maximizing the expression for $t(\{n_i\})$, just as in section 2.1.5.

5.4.1 Fermi–Dirac statistics

We require to maximize t_{FD}, or more conveniently $\ln t_{\mathrm{FD}}$, subject to the usual macrostate conditions

$$\sum_i n_i = N \tag{5.7}$$

and

$$\sum_i n_i \varepsilon_i = U \tag{5.8}$$

As before, the method is to simplify $\ln t$ using Stirling's approximation (Appendix B), and then to find the conditional maximum using the Lagrange method. An outline derivation follows.

Taking logarithms of (5.4) gives

$$\ln t_{\mathrm{FD}} = \sum_i \{g_i \ln g_i - n_i \ln n_i - (g_i - n_i) \ln(g_i - n_i)\}$$

The Lagrange method writes down for the maximum condition

$$\mathrm{d}(\ln t) + \alpha \mathrm{d}(N) + \beta \mathrm{d}(U) = 0 \tag{5.9}$$

Substituting $\ln t = \ln t_{\mathrm{FD}}$ from above together with (5.7) and (5.8) for N and U gives after differentiation

$$\sum_i \{\ln[(g_i - n_i)/n_i] + \alpha + \beta\varepsilon_i\} \mathrm{d}n_i = 0$$

The Lagrange method enables one to remove the summation sign for the specific, but as yet undetermined, values of the multipliers α and β to obtain for the most probable distribution

$$\ln[(g_i - n_i^*)n_i^*] + \alpha + \beta\varepsilon_i = 0$$

Rearranging this expression we find

$$n_i^* = g_i/[\exp(-\alpha - \beta\varepsilon_i + 1]$$

As anticipated the grouping of the states has no explicit influence on the equilibrium occupation per state, and the result may be written in terms of a distribution function

$$f_i = n_i^*/g_i = 1/[\exp(-\alpha - \beta\varepsilon_i + 1] \qquad (5.10a)$$

The distribution function contains no detail about the states except their energies, and so in concord with the whole density of states approximation it can be thought of as effectively a continuous function of the one-particle energy ε_i or simply ε. Hence a useful form of (5.10a) is

$$f_{\mathrm{FD}}(\varepsilon) = 1/[\exp(-\alpha - \beta\varepsilon) + 1] \qquad (5.10b)$$

This is the Fermi–Dirac distribution function.

5.4.2 Bose–Einstein statistics

The derivation of the Bose–Einstein distribution for a boson gas follows analogous lines.

Taking logarithms of (5.5) and using Stirling's approximation gives

$$\ln t_{\mathrm{BE}} = \sum_i \{(n_i + g_i)\ln(n_i + g_i) - n_i \ln n_i - g_i \ln g_i\}$$

Substituting this into (5.9) together with the restrictions (5.7) and (5.8) gives

$$\sum_i \{\ln[(n_i + g_i)/n_i] + \alpha + \beta\varepsilon_i\} \mathrm{d}n_i = 0$$

The form of the equilibrium distribution is again obtained by setting each term of the sum equal to zero

$$\ln[(n_i^* + g_i)/n_i^*] + \alpha + \beta\varepsilon_i = 0$$

which rearranges to give

$$n_i^* = g_i/[\exp(-\alpha - \beta\varepsilon_i) - 1]$$

In terms of the distribution function this becomes

$$f_i = n_i^*/g_i = 1/[\exp(-\alpha - \beta\varepsilon_i) - 1] \tag{5.11a}$$

or

$$f_{\text{BE}}(\varepsilon) = 1/[\exp(-\alpha - \beta\varepsilon) - 1] \tag{5.11b}$$

This is the Bose–Einstein distribution.

5.4.3 Maxwell–Boltzmann statistics

For the dilute (fermion or boson) gas, the procedure may be followed for the third time, starting now with the expression (5.6) for the number t_{MB} of microstates.

We obtain

$$\ln t_{\text{MB}} = \sum_i \{n_i \ln g_i - n_i \ln n_i + n_i\}$$

which using (5.9) with (5.8) and (5.7) gives

$$\sum_i \{\ln[g_i/n_i] + \alpha + \beta\varepsilon_i\} \mathrm{d}n_i = 0$$

Removing the summation for the equilibrium distribution and rearranging now gives

$$n_i^* = g_i \exp(\alpha + \beta\varepsilon_i)$$

The final result for the Maxwell–Boltzmann distribution is therefore

$$f_i = n_i^*/g_i = \exp(\alpha + \beta\varepsilon_i) \tag{5.12a}$$

or

$$f_{\text{MB}}(\varepsilon) = \exp(\alpha + \beta\varepsilon) \tag{5.12b}$$

One may note in passing that this distribution bears a marked similarity to the Boltzmann distribution for localized particles derived in Chapter 2. This greatly simplifies the discussion of dilute gases, since we have unknowingly already covered much of the ground, as we shall see in Chapter 6.

5.4.4 α and β revisited

After the lengthy discussion concerning α and β in Chapter 2, brevity is now in order. The parameter α again is related to the restriction (5.7) which spawned it. In the MB case this involves simply a normalizing constant. In the full FD and BE cases, the general idea is the same (α is adjusted until the distribution contains the correct number N of particles) but the mathematics is not so simple!

For the parameter β, the arguments of section 2.3.1 remain valid. They did not depend on the two assemblies in thermal contact having localized particles. Therefore β must be a common parameter between *any* two assemblies in thermal equilibrium, and following section 2.3.2 we continue to use the identity: $\beta = -1/k_B T$.

The final outcome of this chapter is to write the results of our hard work all in a single composite equation. The distribution function, defined by the average number of particles per state of energy ε_i, is given for a gaseous assembly in thermal equilibrium at temperature T by

$$f_i = \frac{1}{B\exp(\varepsilon_i/k_B T) \quad \begin{matrix} +1 & \text{(FD)} \\ 0 & \text{(MB)} \\ -1 & \text{(BE)} \end{matrix}} \qquad (5.13)$$

In (5.13) the choice of $+1$, 0 or -1 is governed by which of the three types of statistics is relevant. The parameter B (equivalently α since $B = \exp(-\alpha)$) is adjusted to give the correct number of gas particles. How the three distributions are used in practice will emerge from the following four chapters.

5.5 SUMMARY

This chapter builds on previous ideas to derive the distribution functions for gases in thermal equilibrium. We use the statistical approach first outlined in Chapter 1 together with the counting of states of Chapter 4

1. Grouping together g_i of the numerous one-particle states is worthwhile (and does no violence to the physics) since the simplifying approximations of large numbers can then be used.
2. The quantum mechanics of identical particles spawns two classes, bosons and fermions.
3. Half-integral spin particles are fermions; zero or integral spin particles are bosons.
4. Occupation of states for fermions is restricted by the Exclusion Principle – no two fermions can occupy the same state. There is no such restriction for the (friendly!) bosons.
5. Counting microstates in each group reduces to a simple binomial problem (Appendix A).
6. When occupation of states is small (a 'dilute gas'), fermion gases and boson gases both tend to the same limit, the Maxwell–Boltzmann or classical limit.

7. Use of Stirling's approximation and Lagrange multipliers, as used earlier, gives the Fermi–Dirac, Bose–Einstein and Maxwell–Boltzmann distributions, with parameters α and β.
8. α is determined by the particle number, and β relates to temperature as before.
9. The final result for the three distributions (FD, BE and MB) can be given in a single expression (5.13).

6

Maxwell–Boltzmann gases

As a first application of the groundwork of the two previous chapters, we consider the simplest situation. This is a gas for which the Maxwell–Boltzmann (dilute) limit is valid. Furthermore we shall consider only monatomic gases in the present chapter, leaving the complications (and the interest!) of diatomic gases until Chapter 7. First we need to decide the practical range of validity of the MB limit. Next we can derive the MB distribution of speeds in the gas. Finally we may work out the thermodynamic properties of the gas and compare these to results from the ideal gas laws.

6.1 THE VALIDITY OF THE MAXWELL–BOLTZMANN LIMIT

The MB distribution ((5.12) and (5.13)) may be written as

$$f_i = A \exp(-\varepsilon_i / k_B T) \tag{6.1}$$

with the constant $A = \exp(\alpha) = 1/B$. Our statistical method so far (and until Chapter 14) applies only to perfect gases, in the sense of gases whose particles are weakly interacting. But in addition the MB distribution applies only to a gas sufficiently dilute that all the occupation numbers n_i are much less than the number of states g_i, i.e. that all $f_i \ll 1$. Taking the ground state energy as the energy zero (or close to it), the dilute condition therefore means that the constant A in (6.1) should be $\ll 1$.

Clearly in order to explore further we need to calculate A. This is done using its associated condition (5.7):

$$\begin{aligned}
N &= \sum_i n_i \\
&= \sum_i g_i f_i \\
&= A \sum_i g_i \exp(-\varepsilon_i / k_B T) \\
&= AZ
\end{aligned} \tag{6.2}$$

where the *partition function* Z is defined as the sum over all one-particle states of the Boltzmann factors $\exp(-\varepsilon_i/k_BT)$, just as in the Boltzmann statistics of localized particles (section 2.4). Hence the problems of calculating A and Z are identical. To make progress we now need the details of g_i, and that was the topic of Chapter 4.

To be specific, consider a monatomic, zero-spin gas such as ^{4}He. The states of a gas particle are then precisely those of 'fitting waves into boxes', (4.4) or (4.5) with $G = 1$. We have

$$g(k)\delta k = V/(2\pi)^3 \cdot 4\pi k^2 \delta k \qquad (6.3)$$

as the number of states between k and $k + \delta k$, the wavevector k going from 0 to ∞.

Within the density of states approximation, the partition function is then calculated as an integral

$$Z = \int_0^\infty V/(2\pi)^3 \cdot 4\pi k^2 \exp(-\varepsilon(k)/k_BT)\mathrm{d}k \qquad (6.4)$$

The integral may be evaluated in (at least) two ways. One is to transform the density of states from k to energy ε, i.e. to evaluate

$$Z = \int_0^\infty g(\varepsilon)\exp(-\varepsilon/k_BT)\mathrm{d}\varepsilon$$

with $g(\varepsilon)$ precisely that worked out in section 4.3 (see (4.9)). The other entirely equivalent route is to use the dispersion relation $\varepsilon(k) = \hbar^2k^2/2M$, (4.7), to transform the energy in (6.4), leaving a tractable integral over k. Using the latter method, we obtain

$$Z = V/(2\pi)^3 \cdot 4\pi \int_0^\infty k^2 \exp(-bk^2)\mathrm{d}k \qquad (6.5)$$

with $b = \hbar^2/(2Mk_BT) = h^2/(8\pi^2Mk_BT)$. The integral in (6.5) is precisely equal to the standard integral I_2 discussed in Appendix C, so that the expression becomes

$$\begin{aligned} Z &= V/(2\pi)^3 \cdot 4\pi \cdot (I_2/I_0) \cdot I_0 \\ &= V/(2\pi)^3 \cdot 4\pi \cdot (1/2b) \cdot (\pi/4b)^{1/2} \\ &= V(2\pi Mk_BT/h^2)^{3/2} \end{aligned} \qquad (6.6)$$

Equation (6.6) is a central result for the MB gas, and we shall use it later in the chapter to calculate the thermodynamic functions of the gas. Meanwhile we return to the question of the validity of MB statistics. Having calculated Z, we have effectively calculated the constant $A = N/Z$ (equation (6.2)). For MB statistics to be valid we require $A \ll 1$.

It is worth putting in the numbers for the worst possible case! Consider ^{4}He at 1 atmosphere pressure and at a temperature of 5 K, at which it is still a gas (the boiling point is 4.2 K). Substituting the numbers (check it using the necessary constants from Appendix D?) gives

$$A = (N/V) \cdot (h^2/2\pi M k_B T)^{3/2} \tag{6.6a}$$
$$= 0.09$$

for this case.

This is a useful calculation. What it shows is that even for ^{4}He (and similarly for ^{3}He which boils at 3.2 K) the value of A is sufficiently small to justify the use of MB statistics as a first approximation (and as we see later in section 6.3, MB statistics lead to the perfect gas laws). A is often called the *degeneracy parameter*, $A \ll 1$ being the classical or 'non-degenerate' limit.

For helium gas near to its boiling point, the value $A = 0.09$ suggests that degeneracy will be a small but significant cause of deviation from the perfect gas laws. It is not too straightforward to identify experimentally, since just above the boiling point one unsurprisingly finds that corrections due to non-ideality (interactions between atoms, finite size of atoms) also cause significant deviations. However, the degeneracy corrections are particularly interesting; since ^{4}He is a boson and ^{3}He is a fermion, the deviations may be expected to be of opposite signs. And so it is found.

For all other real chemical gases, and for helium at more reasonable temperatures, the value of A is even smaller, since the mass M and temperature T both enter as inverse powers. For example, air at room temperature and pressure has $A \approx 10^{-5}$. The nearest competitor to helium is hydrogen gas, but this boils around 20 K. On the other hand, in the free electron gas model of a metal, one uses an electron gas at the metallic density. Here $A \gg 1$ since the mass is so small, so that the gas is degenerate and FD statistics must be used (see Chapter 8).

To conclude the section, let us note that A and Z are truly quantum quantities. They depend on Planck's constant h, and on the spin factor G (=1 for ^{4}He). But when the numbers are substituted, we find that the dilute MB limit is entirely justified for real gases. Hence, on that basis, the rest of the chapter is worth the effort!

6.2 THE MAXWELL–BOLTZMANN DISTRIBUTION OF SPEEDS

Without really trying, we have in fact derived the distribution of speeds of gas molecules in an ideal gas, the distribution which plays an important part in the kinetic theory of gases.

In kinetic theory, one often requires the number of molecules which have (scalar) velocities between v and $v + \delta v$. This is given directly by the MB distribution – it is as easy as $n_i = g_i \times f_i$. The number we require is conveniently written as $n(v)\delta v$, defining the function $n(v)$ as a density of particles in speed v. Hence we have

$$n(v)\delta v = g(v)\delta v \cdot A \exp(-\varepsilon(v)/k_B T) \tag{6.7}$$

in which $g(v)\delta v$ is defined as the number of states in the range of interest, i.e. with speeds between v and $v + \delta v$.

The number of states is obtained directly from the 'fitting waves into boxes' result, (6.3), with the simple linear transformation: momentum = $\hbar k = Mv$, i.e. $k = (M/\hbar)v$ (remembering that the range δk must be transformed to δv as well as the k^2 term to v^2!). The energy $\varepsilon(v)$ equals $Mv^2/2$; and the constant A is N/Z, with Z given by (6.6). Hence (6.7) is reduced to be a function of v only

$$n(v)\delta v = Cv^2 \exp(-Mv^2/2k_\mathrm{B}T)\delta v \tag{6.8}$$

with

$$C = 4\pi N(M/2\pi k_\mathrm{B}T)^{3/2} \tag{6.8a}$$

This is the required result, obtained originally by Maxwell. It is an entirely classical result, as can be seen by the fact that h appears nowhere in it. In our derivation, the h^{-3} factor in Z (a detailed quantum idea, as discussed above) cancelled with an h^3 from the transformation from k to v. Another way of obtaining the constant C makes this point clear. If (6.8) is to describe the properties of N particles (gas molecules), then it must satisfy the normalization requirement

$$N = \sum_i n_i = \int_0^\infty n(v)\mathrm{d}v$$

Integration of (6.8), having replaced the range δv by $\mathrm{d}v$, may be achieved using the integral I_2 (Appendix C). It will be found that the value of C which satisfies this normalization condition is again given by (6.8a).

More comprehensive discussions of the properties and uses of the speed distribution are found in many books on kinetic theory. Some of its properties are illustrated in Fig. 6.1. The three different but representative speeds indicated on the graph are all of order $(k_\mathrm{B}T/M)^{1/2} = v_T$, say. They are as follows:

1. $v_{\max}$ ($= \sqrt{2}v_T$), the most probable speed corresponding to the maximum of the curve.
2. $\bar{v}$ ($= \sqrt{(8/\pi)}v_T$), the mean speed of the molecules. This is again calculated using the integrals of Appendix C with $b = M/2k_\mathrm{B}T$. We write $\bar{v} = \int_0^\infty vn(v)\mathrm{d}v / \int_0^\infty n(v)\mathrm{d}v = I_3/I_2$, which gives the stated result.
3. v_{rms} ($= \sqrt{3}v_T$), the root mean square speed. This is calculated similarly (albeit with even less difficulty since it involves the recurrence relation of Appendix C only) since the mean square speed $v_{\mathrm{rms}}^2 = I_4/I_2$.

The evaluation of the mean square speed is of particular significance to the thermal properties, since it proves that the average (kinetic) energy per molecule is given by

$$\bar{\varepsilon} = Mv_{\mathrm{rms}}^2/2 = 3k_\mathrm{B}T/2 \tag{6.9}$$

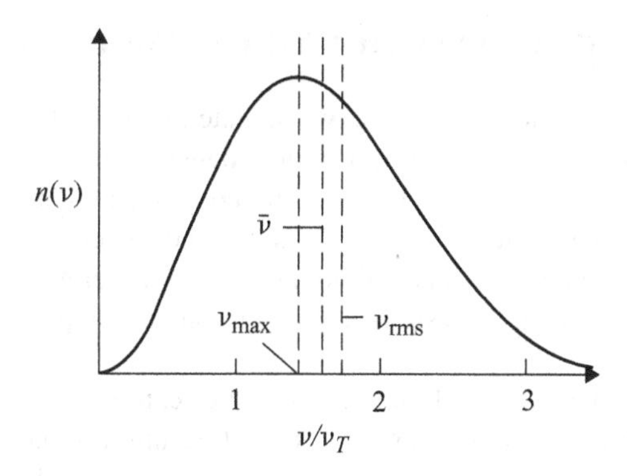

Fig. 6.1 The Maxwell–Boltzmann distribution of speeds in a gas. The representative speeds v_{max}, $\bar{v}$ and v_{rms} are defined in the text.

Also since the gas is isotropic and has no net velocity the three component velocities must satisfy

$$\begin{aligned} Mv_x^2/2 &= Mv_y^2/2 = Mv_z^2/2 = M\,(v_x^2 + v_y^2 + v_z^2)/6 \\ &= Mv_{rms}^2/6 = k_B T/2 \end{aligned}$$

This is another example of the classical *law of equipartition of energy* referred to earlier in section 3.2. Each degree of freedom of the system, i.e. the translational motion of each molecule in each of the three dimensions of space, contributes $\frac{1}{2}k_B T$ to the internal energy of the system. We shall see this result again in the following section and in Chapter 7.

Finally, before leaving the topic, we may note that other more detailed statistical information can be compiled and used in a similar way. Again the key is $n_i = g_i \times f_i$, i.e. number of particles in a group = number of states in the group × filling factor. Suppose for example one wants the number of particles with a velocity close to a particular value $\boldsymbol{v}$. We may define the usual density functions for n and g such that $n(v_x, v_y, v_z)\mathrm{d}v_x\mathrm{d}v_y\mathrm{d}v_z$ is the number of particles and $g(v_x, v_y, v_z)\mathrm{d}v_x\mathrm{d}v_y\mathrm{d}v_z$ the number of states with velocity in the element $\mathrm{d}v_x\mathrm{d}v_y\mathrm{d}v_z$ at velocity $\boldsymbol{v} = (v_x, v_y, v_z)$. Fitting waves into boxes in this case gives the g function to be a constant; the states are evenly spread in k-space, and hence in v-space also. And the Boltzmann factor in f_i is simply $\exp(-Mv^2/2k_B T)$ with $v^2 = v_x^2 + v_y^2 + v_z^2$. Therefore the result is

$$\begin{aligned} &n(v_x, v_y, v_z)\mathrm{d}v_x\mathrm{d}v_y\mathrm{d}v_z \\ &\qquad = \mathrm{const} \times \exp(-Mv^2/2k_B T)\mathrm{d}v_x\mathrm{d}v_y\mathrm{d}v_z \end{aligned} \tag{6.10}$$

with the value of the normalization constant left as a problem (Appendix E).

6.3 THE CONNECTION TO THERMODYNAMICS

This is an important section. We can now calculate the properties of a gas of weakly interacting spinless particles in the limit of low density. These are precisely the conditions for an ideal gas, one which obeys the equation of state $PV = RT$ with T being the thermodynamic temperature. We shall find the comforting result that our gas also obeys this equation of state, although our T was defined statistically from β. Hence we shall fully justify the statistical definitions of temperature and of entropy (see section 2.3).

The calculation methods for Boltzmann statistics can be used again here, so reference should be made back to section 2.5. The only caution is that the expressions for t^* are similar but not identical for the two cases. For localized particles (2.3) has a factor of $N!$, whereas the corresponding (5.6) for a gas of indistinguishable particles does not.

6.3.1 Internal energy and C_V

Since we have gone to the trouble of evaluating Z for a monatomic spinless gas (see (6.6)), the quickest way to U and C_V is to use Method 2 of section 2.5. The derivation of (2.27) depended only on the definition of Z, which is the same for gases as for localized particles. Hence

$$U = N\mathrm{d}(\ln Z)/\mathrm{d}\beta \tag{2.27}$$

$$= Nk_\mathrm{B}T^2\mathrm{d}(\ln Z)/\mathrm{d}T \tag{6.11}$$

using $\beta = -1/k_\mathrm{B}T$. Since $Z \propto T^{3/2}$ (equation (6.6)), (6.11) at once gives

$$U = \tfrac{3}{2}Nk_\mathrm{B}T \tag{6.12}$$

This is the result expected from the kinetic theory treatment of (6.9), since $U = N\bar{\varepsilon}$. The internal energy U for our perfect gas is a function of T only – it does not depend on V; and it is proved to have the value given by the old equipartition law. The expression for C_V follows at once

$$C_V = \mathrm{d}U/\mathrm{d}T = \tfrac{3}{2}Nk_\mathrm{B} \tag{6.13}$$

The heat capacity is a constant in the MB limit – a warning for lovers of the third law of thermodynamics that this limit cannot survive down to the absolute zero of temperature.

6.3.2 Entropy

To find the entropy, we use Method 1 of section 2.5, namely a direct evaluation of S from its statistical definition $S = k_\mathrm{B} \ln \Omega$. For the MB gas, the calculation proceeds as follows:

$$
\begin{aligned}
S/k_{\mathrm{B}} &= \ln \Omega && \text{definition} \\
&= \ln t^* && \text{usual approximation} \\
&= \ln\left(\prod g_i^{n_i}/n_i!\right) && \text{using } t_{\mathrm{MB}}, (5.6) \\
&= \sum (n_i \ln g_i - n_i \ln n_i + n_i) && \text{Stirling's approximation} \\
&= \sum n_i(\ln g_i/n_i + 1) && \text{rearranging} \\
&= \sum n_i(\ln Z - \ln N + \varepsilon_i/k_{\mathrm{B}}T + 1) && \text{MB result for } n_i/g_i, \text{ i.e. (6.1) with } A = N/Z \\
&= N(\ln Z - \ln N + 1) + U/k_{\mathrm{B}}T && \text{identifying } U
\end{aligned}
$$

Hence the result for any MB gas (in this proof, no assumptions are made about any specific monatomic gas etc.) is

$$S = Nk_{\mathrm{B}}(\ln Z - \ln N + 1) + U/T \tag{6.14}$$

For the monatomic, spinless gas under particular consideration in this chapter, we have evaluated Z, (6.6), and U, (6.12). Hence in this case we have

$$S = Nk_{\mathrm{B}}(\ln V - \ln N + \tfrac{3}{2}\ln T) + S_0 \tag{6.15}$$

with the 'entropy constant' S_0 given by

$$S_0 = Nk_{\mathrm{B}}[\tfrac{3}{2}\ln(2\pi M k_{\mathrm{B}}/h^2) + \tfrac{5}{2}] \tag{6.15a}$$

Equation (6.15) (the Sackur–Tetrode equation) is an interesting result. It is classical in the sense that it cannot be correct down to the absolute zero; $\ln T \to -\infty$ whereas the physical S has a lower limit of zero at $T = 0$. Nevertheless it contains Planck's constant in the entropy constant S_0. Furthermore this constant can be checked by experiment as follows. We measure the specific and the latent heats of a specific substance from (essentially) $T = 0$ up to a temperature at which the substance is an ideal gas. This enables us to calculate S calorimetrically, using the fact that $S = 0$ at $T = 0$, from an expression of the type

$$S_{\mathrm{cal}} = \int_0 (C/T)\mathrm{d}T + L_1/T_1 + L_2/T_2$$

the subscript 1 referring to the solid–liquid transition and 2 to the liquid–gas transition. In this way the value (6.15a) of the constant S_0 has been accurately verified.

6.3.3 Free energy and pressure

Now let us turn to the free energy F, and Method 3 of section 2.5, the 'royal route'. Without reference as yet to any specific gas, the hard work was done in the previous section in the derivation of (6.14). One can see that a very simple expression for F emerges, namely

$$F \equiv U - TS = -Nk_{\mathrm{B}}T(\ln Z - \ln N + 1) \tag{6.16a}$$

Using Stirling's approximation in reverse, this may be neatly written

$$F = -Nk_{\mathrm{B}}T \ln Z + k_{\mathrm{B}}T \ln N! \tag{6.16b}$$

The answer is the same as (2.28) for localized particles but with the addition of the $N!$ term. Our arguments from (6.16) branch along three rather distinct lines.

1. *Pressure.* First we can readily use $F(T, V, N)$ to calculate the pressure as follows:

$$P = -(\partial F/\partial V)_{T,N}$$

since

$$\mathrm{d}F = -S\mathrm{d}T - P\mathrm{d}V + \mu\mathrm{d}N$$

Hence

$$P = Nk_{\mathrm{B}}T(\partial \ln Z/\partial V)_{T,N}$$

using (6.16).

For the monatomic MB ideal gas, since $Z \propto V$ (equation (6.6)), this becomes

$$P = Nk_{\mathrm{B}}T/V \tag{6.17}$$

The result (6.17) remains true for polyatomic MB gases, since we shall see in Chapter 7 that $Z \propto V$ for these also; the box volume only enters Z via the translational motion of the molecules. Since this is identical to the ideal gas law, $PV = RT$, this justifies completely our statistical definitions of temperature and entropy. By calculating P and comparing with the ideal gas law we have verified

$$\beta = -1/k_{\mathrm{B}}T \tag{2.20}$$

and

$$S = k_{\mathrm{B}} \ln \Omega \tag{1.5}$$

with the constant $k_{\mathrm{B}} = R/N$, the gas constant per molecule, i.e. Boltzmann's constant.

2. *Extensivity*. Next, a few words about that extra $N!$ in (6.16b). For our treatments of gases or of localized particles to make sense, we require that the extensivity is correct. If we double the amount of the substance, we expect functions like U, S and F also to double, so long as the doubling is done at constant T and at constant density N/V.

   Now for a solid, the equation $F = -Nk_BT \ln Z$, (2.23), satisfies this requirement. The energy levels ε_j depend only on the specific volume, V/N. Hence $Z = \sum \exp(-\varepsilon_j/k_BT)$ is unchanged at given T and N/V. Therefore $F \propto N$ as required.

   However, for a gas the extra $N!$ term is needed. Since all the gas particles are competing for the same space, we have $Z \propto V$, but independent of N, (6.6). It is now Z/N which is unchanged when N is altered at constant density. Therefore (6.16): $F = -Nk_BT(\ln Z/N + 1)$ has the correct properties. The bracket remains unchanged and again $F \propto N$.
3. *The Gibbs paradox*. This was a classical problem of indistinguishability, perhaps a paradox no longer. Consider a box of fixed volume V and temperature T containing a mixture of two ideal gases A and B. We adopt a notation in which the subscript A refer to the properties which gas A would have in the *absence* of gas B; and similarly for subscript B. If the two gases are different then they behave independently, they occupy different states. Hence $\Omega = \Omega_A \times \Omega_B$, and $S = S_A + S_B$, $F = F_A + F_B$, $P = P_A + P_B$, etc. The two gases behave as if the other were not present. Even an isotopic mixture, say of ^{3}He and ^{4}He, behaves in this way.

   However, the situation is different if the two gases A and B are identical. It is true that $P = 2P_A$, but $S \neq 2S_A$, and $F \neq 2F_A$. The molecules are now competing for states, so the statistical properties of the second gas are modified by the existence of the first. In fact we can see from (6.14) that $S = 2S_A - Nk_B \ln 2$ (arising from the $N \ln N$ term, i.e. from the $N!$) where N is the total number of A molecules. The degree of disorder is lessened by the competition.

6.4 SUMMARY

This chapter derives the properties of an ideal monatomic gas in the dilute limit.

1. The dilute limit is found to be a valid approximation for all real chemical gases.
2. The partition function Z, summing Boltzmann factors $(\exp(-\varepsilon_i/k_BT))$ over all states, again plays a useful role.
3. MB statistics leads directly to the speed distribution of gas molecules, first derived by Maxwell.
4. The MB gas is shown to have the equation of state $PV = RT$.
5. Our statistical temperature (based on $\beta = -1/k_BT$) is thus identical to the thermodynamic Kelvin temperature.
6. The kinetic energy of the gas molecules in thermal equilibrium gives an illustration of the classical principle of equipartition of energy.

7. However, it is evident from expressions for S and C that classical ideas must fail at low enough temperatures.
8. Surprisingly, calorimetric methods can determine Planck's constant through the entropy constant S_0 (6.15).
9. MB statistics differs from Boltzmann statistics by having a (necessary) extra $N!$ term in the free energy expression (6.16b). This arises from the indistiguishability of gaseous particles, a matter of controversy in Boltzmann's day.

7

Diatomic gases

This chapter is a slight diversion, and could well be omitted at a first reading. However, the study of diatomic Maxwell–Boltzmann gases proves to be a rather interesting one. It will reinforce the ideas of energy scales, introduced in Chapter 2, and illustrate further the concept of degrees of freedom. Furthermore the rotation of molecular hydrogen (H_2) gas holds a few quantum surprises. Throughout the chapter we shall assume, as is realistic, that MB statistics applies. The quantum surprises are not concerned with 'degeneracy', i.e. whether FD or BE corrections need be made to the statistics. Rather they are to do with the indistinguishability of the two H atoms which make up the H_2 molecule.

7.1 ENERGY CONTRIBUTIONS IN DIATOMIC GASES

As outlined in the previous chapter, if the partition function of our MB gas is evaluated then all its thermal properties can be calculated. So far we have treated explicitly only a monatomic gas such as helium. What now about a polyatomic gas?

The problem is quite tractable, with one basic assumption. This is that the various forms of possessing energy are independent. To explain what this means, consider the contributions to the energy of a polyatomic gas molecule. The molecule can have energy due to translational motion, due to rotation, due to internal vibrations and (exceptionally) due to electronic excitation. If these energy contributions are independent, then it means, for example, that the state of vibration does not influence the possible energies of translation. In other words the one-particle states can be described simply as having energies

$$\varepsilon = \varepsilon_{\text{trans}} + \varepsilon_{\text{rot}} + \varepsilon_{\text{vib}} + \varepsilon_{\text{elec}} \tag{7.1}$$

so that each mode of possessing energy can be considered separately from the others.

This is an approximation only, but it is a good one. Translation, i.e. motion of the centre of mass, is accurately independent of the internal degrees of freedom of the molecule, i.e. motion around the centre of mass and excited electronic states. One might expect a small coupling between vibration and rotational states; for example, in a state of high rotational energy of a diatomic molecule the bond will stretch a little from the inertia of the atoms, and this will influence the bond strength and hence the vibrational frequency of the molecule. However, such effects are small in practice and may be (gratefully) neglected.

From the point of statistical physics, this independence has a great simplifying influence. It means that the partition function Z factorizes.

The partition function is defined (equation (6.2)) as the sum over all states of the Boltzmann factors of every state. Writing β for $-1/k_B T$, and using (7.1) we obtain

$$\begin{aligned} Z &= \sum_{\text{all states}} \exp[\beta(\varepsilon_{\text{trans}} + \varepsilon_{\text{rot}} + \varepsilon_{\text{vib}} + \varepsilon_{\text{elec}})] \\ &= \sum_{\text{trans}} \exp[\beta(\varepsilon_{\text{trans}})] \times \sum_{\text{rot}} \exp[\beta(\varepsilon_{\text{rot}})] \\ &\quad \times \sum_{\text{vib}} \exp[\beta(\varepsilon_{\text{vib}})] \times \sum_{\text{elec}} \exp[\beta(\varepsilon_{\text{elec}})] \qquad (7.2) \\ &= Z_{\text{trans}} \times Z_{\text{rot}} \times Z_{\text{vib}} \times Z_{\text{elec}} \qquad (7.3) \end{aligned}$$

The simplicity of (7.2) is that each full state of the molecule, by the assumption of independence, can be specified by its separate quantum numbers (i.e. state labels) for translation, rotation, vibration and electronic excitation. Hence the partition function factorizes as in (7.3) into independent component parts.

Since Z factorizes, $\ln Z$ has a number of additive terms. As a result there are independent additive terms in the thermodynamic functions such as F (equation (6.16)), U (equation (6.12)) and hence C_V. It is particularly instructive to note how this works out for the free energy F, since it becomes clear how to handle the $\ln N!$ term of (6.16b).

Substituting the form (7.3) for the diatomic gas into the general expression (6.16b) for any MB gas, we obtain

$$\begin{aligned} F &= -Nk_B T \ln Z + k_B T \ln N! \\ &= -Nk_B T \ln Z_{\text{trans}} + k_B T \ln N! - Nk_B T \ln Z_{\text{rot}} \\ &\quad - Nk_B T \ln Z_{\text{vib}} - Nk_B T \ln Z_{\text{elec}} \qquad (7.4) \\ &= F_{\text{trans}} + F_{\text{rot}} + F_{\text{vib}} + F_{\text{elec}} \qquad (7.5) \end{aligned}$$

The free energy of the gas is decomposed as anticipated into various parts. The translational part F_{trans} is defined by the first two terms of (7.4), which includes Z_{trans} together with the $\ln N!$ term. Note that this is identical to the total free energy of a gas of structureless (i.e. monatomic) molecules as worked out in Chapter 6. Hence F for the diatomic gas is equal to F for the monatomic gas *plus* additive contributions from the internal degrees of freedom, the final three terms of (7.4). The extra contribution of each internal degree of freedom is particularly simple. It gives merely a $-Nk_BT \ln Z$ addition to F; and this is identical to the free energy of a set of *localized* particles (Chapter 2) with appropriate one-particle states. Therefore, we shall find that most of the hard groundwork for this chapter is already done, so that we may concentrate on the results.

7.2 HEAT CAPACITY OF A DIATOMIC GAS

To be specific, let us use the ideas of the previous section to calculate the heat capacity C_V of a typical diatomic gas.

7.2.1 Translational contribution

As already indicated, this term holds nothing new above the content of Chapter 6. The translational motion of any molecule is described by the k-states of 'fitting waves into boxes' fame. And this treatment leads to a partition function, (6.6)

$$Z_{\text{trans}} = V(2\pi M k_B T/h^2)^{3/2} \tag{7.6}$$

with M being the mass of the molecule. Hence as before $U_{\text{trans}} = \frac{3}{2}Nk_BT$ and the heat capacity contribution is given by

$$C_{V,\text{trans}} = \frac{3}{2}Nk_B$$

7.2.2 Electronic contribution

When we come to the internal contributions, (7.4) and (7.5) tell us that we can forget about the molecules being specifically those of a gas. Rather we can treat the contributions as identical to those from an assembly of distinguishable particles.

Allowance for thermal excitation into a higher electronic state of energy ε above the ground state is identical to the treatment of the two-state system of section 3.1. The result in the heat capacity is to add to C_V a 'Schottky anomaly' term as given in (3.4) and Fig. 3.3.

In practice this contribution to C_V is negligible in almost all cases. This is because typical values of ε are around 10 eV, so that we are in the extreme quantum limit ($k_BT \ll \varepsilon$) at any realistic temperature. Hence $C_{V,\text{elec}} \approx 0$. The only exceptional everyday gases are NO and O_2. In these ε happens to be particularly small and the Schottky anomaly peaks are calculated to appear around 75 K for NO and 5000 K for O_2. However, with these exceptions it is justifiable to neglect the electronic term, and we shall do so in the rest of this chapter.

7.2.3 Vibrational contribution

Similarly, the contribution from vibrational motion has been calculated earlier in this book. Each diatomic molecule is accurately represented by a one-dimensional harmonic oscillator, and furthermore the molecules are very weakly interacting. Therefore, the vibrations of our N gas molecules can be described as an assembly of N identical and weakly interacting harmonic oscillators. This is the problem discussed fully in section 3.2. The heat capacity contribution $C_{V,\mathrm{vib}}$ is given precisely by (3.15) and is illustrated in Fig. 3.10.

The heat capacity $C_{V,\mathrm{vib}}$ is calculated to rise from zero to Nk_B as the temperature is raised, the change occurring around the scale temperature θ. As discussed in section 3.2, $k_\mathrm{B}\theta$ is the energy level spacing of an oscillator, equal to $h\nu$ where ν is the classical frequency. Its value, therefore, will depend on the bond strength between the two atoms of the molecule and on their masses. For typical common gases θ turns out to have a value in the 2000–6000 K region (roughly 2000 K for O_2, 3000 K for N_2 and 6000 K for H_2 where the masses are lighter). At room temperature, therefore, vibration remains substantially unexcited so that $C_{V,\mathrm{vib}} \approx 0$ for all diatomic gases. At elevated temperatures the onset of vibration is seen, and the heat capacity rises. However, in practice, it often happens that, before $C_{V,\mathrm{vib}}$ becomes very large, the diatomic gas dissociates into $2N$ monatomic gas molecules, the vibrational energy having overcome the bonding energy.

7.2.4 Rotational contribution

Having effectively disposed of electronic excitation and vibration, both theoretically and experimentally, we now turn to the rotational contribution. This will be found a more substantial topic!

The rotation of a linear molecule is modelled, bearing in mind the strictures of section 7.1, by the motion of a rigid rotator. The rotator has a fixed moment of inertia I around an axis perpendicular to its own axis; spinning motion around its own axis is neglected. The solution for the quantum states of the rotator will be quoted here, but they should look familiar to a student who has studied any angular momentum problem in quantum mechanics. Basically the requirement for the wavefunction to be single valued upon a 2π rotation leads to quantization of an angular momentum component in units of $\hbar$. The allowed values of (angular momentum)^2 become $l(l+1)\hbar^2$, with $l = 0, 1, 2, \ldots$. And hence the allowed energy levels are given by

$$\begin{aligned} \varepsilon_l &= l(l+1)\hbar^2/2I \\ &\equiv l(l+1)k_\mathrm{B}\Phi \end{aligned} \tag{7.7a}$$

where the temperature Φ so defined represents a characteristic scale temperature for rotation. These levels are degenerate, there being

$$g_l = (2l+1) \tag{7.7b}$$

states to level l, corresponding intuitively to different possible directions for the angular momentum.

Armed with these results we can at once write down the partition function for rotation

$$Z_{\text{rot}} = \sum_{l=0,1\ldots} (2l+1)\exp[-l(l+1)\Phi/T] \tag{7.8}$$

Hence the problem is in principle solved, even if the sum (7.8) must be evaluated numerically rather than in terms of known functions.

Moreover, for all gases except hydrogen, there is a great simplification. When the numbers for the moments of inertia of a gas such as O_2 or N_2 are substituted into (7.7a), we find that the characteristic temperature Φ is about 2 K (2.1 K for O_2, 2.9 K for N_2 but 85 K for H_2). Since these other gases have liquefied (indeed solidified) at such low temperatures, it is always true in the gaseous state that $T \gg \Phi$. Very many rotational states are therefore excited, and the sum (7.8) may be replaced by an integral in this 'classical' limit. We write $y = l(l+1)$ and treat y as a continuous variable, so that $\mathrm{d}y = (2l+1)\mathrm{d}l$. Hence in this approximation

$$\begin{aligned} Z_{\text{rot}} &= \int_0^\infty \exp(-y\Phi/T)\mathrm{d}y \\ &= (-T/\Phi)[\exp(-y\Phi/T)]_0^\infty \\ &= T/\Phi \end{aligned} \tag{7.9}$$

This simple result at once tells us the thermodynamic properties of these diatomic gases. The thermal part of the total partition function Z (equation (7.3)) is obtained by multiplying Z_{trans} (equation (6.6)) by Z_{rot} (equation (7.9)). As discussed in Chapter 3, the ground state contributions from vibration and electronic excitation give rise to zero-point terms only. Since $Z \propto V$ (from the translational contribution only), it remains true that $PV = Nk_BT$ as discussed in section 6.3.3. But the additional factor of T in the partition function from rotation gives an additional Nk_B to C_V. And the total heat capacity is

$$\begin{aligned} C_V &= C_{V,\text{trans}} + C_{V,\text{rot}} \\ &= \left(\frac{3}{2}+1\right)Nk_B \\ &= \frac{5}{2}Nk_B \end{aligned} \tag{7.10}$$

a result which follows from $Z \propto T^{5/2}$, as the reader may verify (compare section 6.3.1). Hence also the ratio $C_P/C_V = \frac{7}{5}$ for the diatomic gas, a ratio in good agreement with experiment.

7.3 THE HEAT CAPACITY OF HYDROGEN

As an example of the above results, we consider in greater detail the heat capacity of hydrogen. From what we have learned so far in this chapter, we anticipate the heat capacity of H_2 to resemble the sketch shown in Fig. 7.1. And in broad terms this is correct.

At temperatures just above the boiling point of H_2, the measured value of C_V is $\frac{3}{2}Nk_B$. At these temperatures, the molecules do not rotate! The moment of inertia of H_2 is so small that the $l = 1$ rotational state is further than k_BT above the ground ($l = 0$) rotational state. In pictorial terms this means that the molecules remain in a fixed orientation as they move in the gas; collisions do not impart enough energy to start rotation. To return to the ideas of Chapter 3, the degree of freedom is frozen out since we are in the extreme quantum limit, i.e. $k_BT \ll \Delta\varepsilon$, where $\Delta\varepsilon$ is the energy level spacing.

Incidentally, we may note in passing that it is even more valid to neglect rotation in a monatomic gas (or axial rotation in the diatomic case). Although the relevant moment of inertia is not zero, it is extremely small since most of the atomic mass is in the nucleus. Hence the first rotational excited state becomes unattainably high, and rotation is frozen out at all temperatures.

At low temperatures, then, only the translational degrees of freedom of the gas are excited, and for the MB gas (equation (6.9)) each molecule contributes $\frac{3}{2}k_BT$ to the total energy from its translation in three dimensions.

By the time room temperature is reached, the rotational motion has become fully excited. Another way of looking at the problem is to note that the rotation of an axially symmetric molecule provides a further two degrees of freedom, two since two angles are needed to specify the direction of the molecule. And at room temperature the classical limit, $k_BT \gg \Delta\varepsilon$, is valid. Hence each molecule now contributes from its

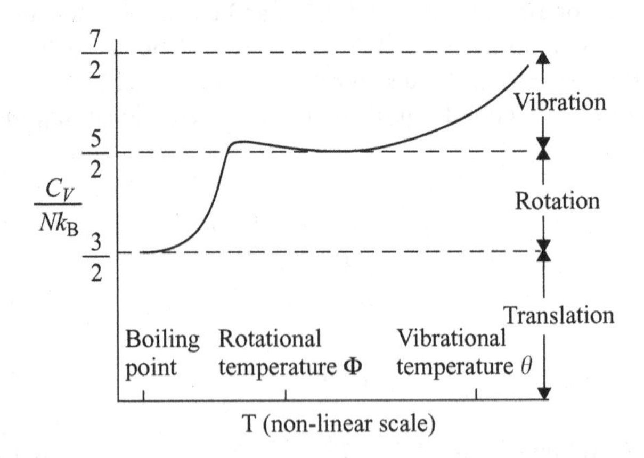

Fig. 7.1 The variation of heat capacity C_V with temperature for a diatomic gas, showing schematically the contributions of translation, rotation and vibration.

five degrees of freedom a total of $\frac{5}{2}k_B T$ to U, and therefore $C_V = \frac{5}{2}Nk_B$, in agreement with (7.10).

At room temperature vibration is in the extreme quantum limit ($k_B T \ll \Delta\varepsilon$) since the characteristic temperature of vibration (around 6000 K) is so high. However, at elevated temperatures, one expects the heat capacity to rise towards $\frac{7}{2}Nk_B$ as vibration becomes excited. This is observed, although the gas dissociates before the classical limit is reached. (The detailed comparison with experiment is also complicated by significant coupling between vibrational and rotational states.)

7.3.1 The onset of rotation

Although the treatment so far gives a satisfactory outline of the properties of hydrogen, the details of the onset of rotation around 50–200 K are not well described. If we use the rotational partition function of (7.8) to calculate $C_{V,\text{rot}}$, we obtain curve A of Fig. 7.2. This is at variance with the experimental results for H_2, which are more like curve C.

Why the discrepancy? Actually an experimenter can help here. Nature is kind in that there are several sorts of 'hydrogen'. Using isotopic separation, it is possible to deduce C_V for the gases H_2, HD and D_2. Even allowing for the different values of Φ (H_2 85 K, HD 64 K, D_2 43 K) arising from the different moments of inertia, the curves for C_V differ markedly. And in fact our theoretical curve A is in good agreement with experiment for HD.

There is a great difference between a heteronuclear and a homonuclear molecule. It is a matter of the identity of the two nuclei. If the nuclei are different, and therefore distinguishable, then what we have done so far is correct. Hence the agreement for HD.

At first sight the modification to Z for identical nuclei might seem to be a simple one, merely to allow for the fact that a rotation of π rather than 2π leads to identity.

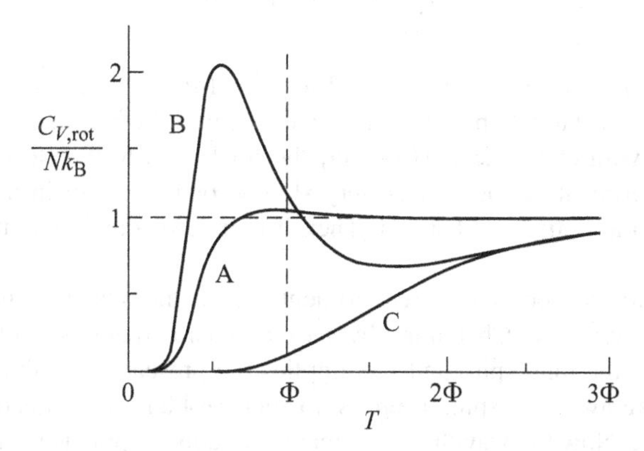

Fig. 7.2 The rotational heat capacity of a diatomic gas. The three curves A, B and C are explained in the text.

However, the truth is more interesting than that, and the result is different for H_2 and for D_2. Let us concentrate upon the common gas H_2. The nuclei are just protons having spin $\frac{1}{2}$. The nuclei are thus identical fermions, and they must therefore be described by an *antisymmetric* wavefunction. In other words the wavefunction must change sign if the co-ordinates of the two protons are interchanged. The following discussion bears similarity to our earlier treatment of bosons and fermions in section 5.2.

The total wavefunction can be decomposed into the product of a 'space part' and a 'spin part'. The space part describes the rotation, and it is found that the $l = 0, 2, 4\ldots$ states have even interchange parity whereas the $l = 1, 3\ldots$ states are odd. In the notation of Chapter 5, the even l states are S (even, symmetric) whereas the odd l states are A (odd, antisymmetric).

Consider first the even l rotational states. Since these are S, the only way to obtain a total wavefunction which is A is for the spin part to be A. And this means that the two spins must be opposite. There is just one such spin state corresponding to the total spin $S = 0$. Hence the even l rotational states must be associated with the spin singlet configuration only, and hence they have spin weighting 1. These states are often called 'para-hydrogen'.

On the other hand odd l states are A, the total wavefunction for H_2 is to be A, and therefore the associated spin part must now be S. This indicates the parallel spin states ($S = 1$), in which there are three possible alignments. Hence the odd l rotational states are linked with the spin triplet states, and they have a weighting of 3. The states are called 'ortho-hydrogen'.

The implication is clear. Assuming that our H_2 is in thermal equilibrium, the partition function of (7.8) (although correct for HD) is in error. Rather than summing equally over the ortho- and para-states, we should have

$$Z_{\text{rot}} = \sum_{\text{even}} (2l + 1) \exp[-l(l + 1)\Phi/T] + 3 \sum_{\text{odd}} (2l + 1) \exp[-l(l + 1)\Phi/T] \tag{7.11}$$

where the even sum goes over $l = 0, 2, 4\ldots$ and the odd sum over $l = 1, 3\ldots$ When (7.11) is used to compute C_V, the result is curve B of Fig. 7.2, in even greater disagreement with experiment. However, the final twist to the story is uncovered when the experiment is done either very slowly, or better still in the presence of a catalyst such as activated charcoal. Then it is indeed found that curve B fits the experiment.

The reason for the poor fit to the experiment without the catalyst is that 'ortho–para conversion' is very slow. Changing l by ± 1 on collision requires a collision process which changes the total spin, and this implies the presence of a third body (i.e. a surface) to take away the spin. There is no such problem for collisions in which l changes by ± 2. Now the way the experiments are done in practice is to prepare and store the H_2 gas at room temperature, i.e. when $T \gg \Phi$. In this limit, the odd and even sums in equation (7.11) are equal (compare the discussion leading to equation (7.9)).

Hence the ratio of ortho- to para-hydrogen molecules when the gas equilibrates at high temperature is 3:1. When the hydrogen is then cooled, and experiments are performed with no ortho–para conversion, then the result is like a 3:1 mixture of two dissimilar gases, and we find that

$$C_V = 0.75\, C_{V,\text{ortho}} + 0.25\, C_{V,\text{para}} \tag{7.12}$$

where the ortho and para heat capacities are calculated respectively only from the odd and even terms of the partition function. This result gives curve C, at last in agreement with the usual experiment.

Two final points. Ortho–para conversion is a technical problem in H_2 in its use in refrigeration. If it is left to equilibrate at low temperatures, then much heat will be released as the $l = 1$ states gradually relax to $l = 0$. In addition all the thermal properties, even the boiling point, change with the ortho:para ratio. Hydrogen is best avoided in cryogenics! The second point is that similar effects occur in the gas D_2, although the details are quite different from H_2 since the deuteron is a boson (S wavefunction) rather than a fermion. We leave this topic as an exercise.

7.4 SUMMARY

In this chapter we stop to examine the thermal properties of diatomic gases, and see that this brings together results from several earlier chapters.

1. A diatomic gas molecule can be considered to have independent energy contributions from translation (as a monatomic molecule), rotation, vibration and electronic excitation.
2. The contribution to the thermal properties (notably to F and C_V) from translational motion is identical to that of a monatomic gas.
3. The other contributions are identical to those derived from Boltzmann statistics of similar localized particles.
4. The partition function, summing Boltzmann factors ($\exp(-\varepsilon_i/k_B T)$) over all states, again plays a useful role.
5. This topic gives a good example of energy scales and the excitation of degrees of freedom (equipartition of energy at high T, quantum 'freezing out' at low T).
6. In practice, electronic excitation and vibration play a minor role for almost all gases.
7. Two-dimensional rotation is fully excited in most diatomic gases, to give $C_V = \frac{5}{2} N k_B$.
8. The onset of rotation is seen in H_2, but it holds additional quantum surprises because the two atoms in the rotating molecule are identical and therefore indistinguishable. Nuclear spin affects the weighting (and sometimes the accessibility) of the odd and even rotational states, and hence the thermal properties.

8

Fermi–Dirac gases

We return now to the main stream of the book, and to the basic statistical properties of ideal gases as introduced in Chapter 5. Of the three types of statistics we have so far discussed only the classical limit, corresponding to Maxwell–Boltzmann gases. In the next two chapters so-called 'quantum statistics' is discussed, that is statistics where the antisymmetric (Fermi–Dirac) or the symmetric (Bose–Einstein) nature of the wavefunction plays a significant role.

The importance of the FD or BE nature is most readily seen when we consider the state of the gas at $T = 0$. At the other extreme, we have already noted that MB statistics is a high-temperature approximation, corresponding to the degeneracy parameter A (defined by (6.6a)) being $\ll$1. As T approaches zero, and therefore A becomes infinitely large, the quantum limit is obvious. For BE statistics, in which any number of particles can occupy a state, the $T = 0$ state is for all N particles to occupy the ground state. This gives a situation of lowest energy and of zero entropy, i.e. of perfect order. In contrast to the 'friendly' bosons, the 'unfriendly' fermions operate an exclusion principle. Therefore the $T = 0$ state for FD statistics is with the N particles neatly and separately packed into the N states of lowest energy, giving a large zero-point energy, but again zero entropy (because of the lack of ambiguity in the arrangement).

The condition $A \ll 1$ for validity of the MB approximation contains other indications about when quantum statistics should be used. First, the value of T needed to make $A = 1$ gives a 'scale temperature' below which quantum effects will dominate, giving estimates of the characteristic temperatures (or equivalent energies) to be introduced in the next two chapters. Second, if we consider $k_B T$ as a thermal energy scale, we can work out a thermal momentum scale and hence (using $p = h/\lambda$) a 'thermal de Broglie wavelength', λ, to characterise the quantum properties of a typical gas molecule. A little rearrangement shows that $A \ll 1$ then translates into $\lambda \ll (V/N)^{1/3}$, which is the average distance between gas particles. So we can see that quantum effects become important when the de Broglie wavelengths of nearby particles overlap, a pretty idea.

In this chapter we now discuss the properties of FD ideal gases. The BE case is treated in Chapter 9.

8.1 PROPERTIES OF AN IDEAL FERMI–DIRAC GAS

As we have seen, FD statistics is needed, (i) when we are dealing with a gas of weakly interacting particles having spin $\frac{1}{2}$ (or $\frac{3}{2}, \frac{5}{2} \ldots$), and (ii) when the gas has a degeneracy parameter $A > 1$. Since

$$A = (N/V)(h^2/2\pi M k_B T)^{3/2} \qquad \text{(8.1) and (6.6a)}$$

one finds in practice that FD statistics is needed only in a few cases of high density N/V, of low temperature T or of low mass M. Important applications are:

1. Conduction electrons in metals at all reasonable temperatures, and also in semiconductors with a high enough carrier density N/V.
2. Liquid ^{3}He at low temperatures.
3. Dense-matter problems in astrophysics, such as in neutron and white dwarf stars.

Since each of these involves spin-$\frac{1}{2}$ particles, we shall explicitly consider spin-$\frac{1}{2}$ fermions only in this chapter. The generalization is straightforward.

8.1.1 The Fermi–Dirac distribution

The distribution function was derived in Chapter 5. The result ((5.10) and (5.13)) is usually written as

$$f_{FD}(\varepsilon) = 1/\{\exp[(\varepsilon - \mu)/k_B T] + 1\} \qquad (8.2)$$

For the rest of this chapter, the subscript FD will be omitted in the interests of clarity. In (8.2), we have made the identification of the 'Fermi energy', μ. This is simply another way of characterizing the parameters α (equation (5.10)) or B (equation (5.13)) as $B = \exp(-\alpha) = \exp(-\mu/k_B T)$. The symbol μ is appropriate since it turns out that this quantity is precisely the same as the chemical potential of the gas.

The form of (8.2) is shown in Fig. 8.1, where it is plotted at three different temperatures. It is not hard to understand. The distribution function $f(\varepsilon)$ is defined as the number of particles per state of energy ε in thermal equilibrium. Equation (8.2) bears the mark of the exclusion principle, since it guarantees that $f \leq 1$ from the $+1$ in the denominator. So curve 1 in the figure, corresponding to a low temperature, shows as expected that $f = 1$ for states with $\varepsilon < \mu$ but $f = 0$ for $\varepsilon > \mu$. There is a change-over region of energy width about $k_B T$ around μ in which f changes from 1 to 0. As T is raised somewhat (curve 2), this change-over region gets wider, although the extremes of the distribution are virtually unaltered. When T is raised further (curve 3) the whole form of the distribution is affected, tending towards the simple exponential MB distribution in the high T limit.

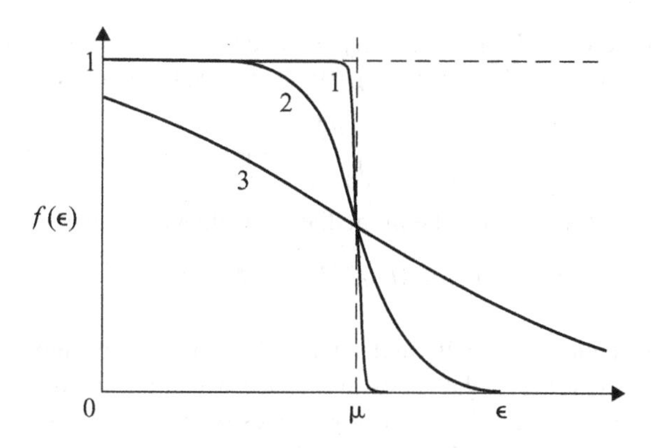

Fig. 8.1 The Fermi–Dirac distribution function at three different temperatures. Curve 1, $k_BT = 0.02\mu$. Curve 2, $k_BT = 0.1\mu$, Curve 3, $k_BT = 0.5\mu$.

8.1.2 The Fermi energy

To use the FD distribution (8.2), we need to know the Fermi energy μ. This parameter (being related to α) is fixed by the number condition $N = \sum n_i$, as explained in section 5.4.4. The value of μ will depend on the macrostate conditions (N, V, T), and in particular it will be a function of T. We start by calculating its value at $T = 0$. There are two obvious ways to proceed, and it is worth being aware of both of them.

Method 1. Use the density of states $g(\varepsilon)$. The definition and the form of $g(\varepsilon)$ should be well-known by this stage. The states are described by fitting waves into boxes, as in Chapter 4, to give the states in k. A transformation using the dispersion relation is then used to give the density of states in ε. The procedure is almost identical to that given in section 4.3 for helium gas, leading to the result (4.9). The only modifications to (4.9) are, (i) that the mass M should refer not to He but to the relevant particle mass, and (ii) that a spin factor $G = 2$ for the spin-$\frac{1}{2}$ fermions should multiply the result. Hence

$$g(\varepsilon)\delta\varepsilon = V4\pi(2M/h^2)^{3/2}\varepsilon^{1/2}\delta\varepsilon \tag{8.3}$$

The determination of $\mu(0)$, the Fermi energy at $T = 0$, follows from the fixed number N of particles in the macrostate. In the density of states approximation we have (directly from the definitions of $g(\varepsilon)$ and of the filling factor $f(\varepsilon)$)

$$N = \int_0^\infty g(\varepsilon)f(\varepsilon)\mathrm{d}\varepsilon \tag{8.4}$$

At the absolute zero, the Fermi function $f(\varepsilon)$ takes the simple form noted earlier, that $f = 1$ for $\varepsilon < \mu(0)$ but $f = 0$ for $\varepsilon > \mu(0)$. Hence (8.4) becomes

$$N = \int_0^{\mu(0)} g(\varepsilon)\mathrm{d}\varepsilon$$

which using (8.3) for $g(\varepsilon)$ may be immediately evaluated to give

$$\mu(0) = (\hbar^2/2M)(3\pi^2 N/V)^{2/3} \tag{8.5}$$

Method 2. Someone who really understands the fitting waves into boxes ideas can use a pretty shortcut here. The states in k-space which are filled at $T = 0$ can be represented as in Fig. 8.2. The low energy states with energy less than $\mu(0)$ are all filled, and since $\varepsilon = \hbar^2 k^2/2M$ these filled states correspond to those with k less than some value (called the Fermi wavevector k_{F}) corresponding to $\mu(0)$. States with $k > k_{\mathrm{F}}$ are unfilled. The sphere of radius k_{F}, representing the sharp boundary between filled and unfilled states, is called the Fermi surface.

The second approach to the determination of $\mu(0)$ is to recognize that this Fermi surface must contain just the correct number N of states. Hence we must have

$$N = V/(2\pi)^3 \times 4\pi k_{\mathrm{F}}^3/3 \times 2 \tag{8.6}$$

where the first factor is the basic density of k-states in k-space, the second is the appropriate volume in k-space (that contained by the Fermi surface), and the final 2 is the spin factor for spin $\frac{1}{2}$. From (8.6) it follows that

$$k_{\mathrm{F}} = (3\pi^2 N/V)^{1/3}$$

and therefore $\mu(0) = \hbar^2 k_{\mathrm{F}}^2/2M$ is precisely as given above in (8.5).

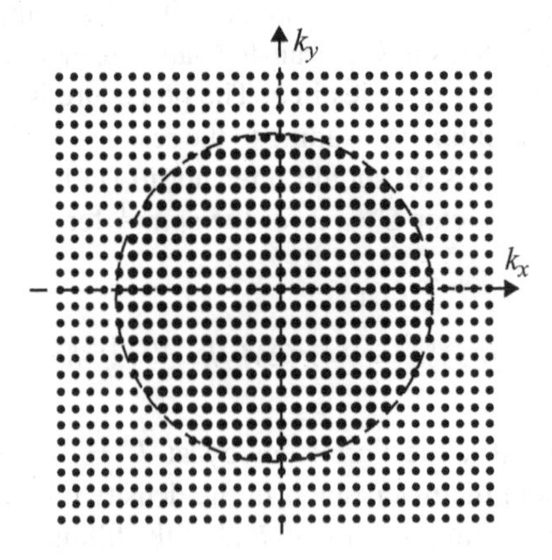

Fig. 8.2 The Fermi surface. Illustrating the occupation of states in k-space at $T = 0$; the sphere has radius k_{F} (• = occupied; • = unoccupied).

Three remarks about the Fermi energy follow.

1. We can now quantify the need for FD statistics in the systems mentioned earlier, by substituting into (8.5). It is useful in so doing to calculate the 'Fermi temperature T_F', defined by:

$$\mu(0) = k_B T_F$$

For example, if we are interested in electrons, (8.5) can be written

$$T_F = 4.1 \times 10^{-15} (N/V)^{2/3}$$

For electrons at metallic densities (say 6×10^{23} electrons in a molar volume of $9 \times 10^{-6}\,\text{m}^3$) this gives a Fermi temperature of about 70 000 K. Therefore, at ambient temperatures we are always in the limit $T \ll T_F$, so that the system is dominated by FD statistics (see again Fig. 8.1). For electrons in a semiconductor, the Fermi temperature will be lower and there are some situations in which MB statistics are adequate. Finally we may note that the dense-free electrons existing in a typical white-dwarf star have Fermi temperatures of around 10^9–10^{10} K. Since the internal temperature of such a star is (only!) a mere 10^7 K, this is again a highly degenerate fermion gas. We return to this topic in Chapter 15.
2. The calculation of $\mu(0)$ can be directly related to the degeneracy parameter A of (8.1). In fact substituting (8.5), together with the definition above of T_F, we obtain

$$A = (8/3\sqrt{\pi})(T_F/T)^{3/2} \approx 1.50(T_F/T)^{3/2}$$

This result explicitly demonstrates that the degeneracy condition $A > 1$ is effectively equivalent to $T < T_F$. And correspondingly the classical limit $A \ll 1$ is the same as $T \gg T_F$.
3. There is a temperature variation to μ. In fact in the classical region μ diverges to minus infinity as $-T \ln T$, a result derivable from the methods of Chapter 6. However, in the degenerate region ($T \ll T_F$) the variation is small, and can often be neglected. It is still determined by (8.4), the number restriction. The temperature enters only through the Fermi function $f(\varepsilon)$. But in the degenerate region (see again curves 1 and 2 of Fig. 8.1), the variation of $f(\varepsilon)$ is only a subtle blurring in an energy range of order $k_B T$ around μ. In fact if $g(\varepsilon)$ were a constant independent of ε, there would be virtually no variation of μ with T in the degenerate region. As T is raised, the filling of states above the Fermi energy would be compensated by the emptying of those below. It is only when $g(\varepsilon)$ varies with ε that a small shift in μ occurs, to allow for the imbalance in numbers of states above and below μ. For a two-dimensional electron gas (exercise!), $g(\varepsilon)$ is a constant and $\mu(T) = \mu(0)$ to a high degree of accuracy. For the usual three-dimensional gas we have a rising

function $g(\varepsilon) \propto \varepsilon^{1/2}$ and μ falls slightly. Using the results of Appendix C it can be shown that for this case

$$\mu(T) = \mu(0)[1 - (\pi^2/12)(k_B T/\mu)^2 \ldots]$$

Since the variation is small, we shall in future not distinguish between μ and $\mu(0)$, unless confusion would arise. Note that the symbol ε_F for the Fermi energy will be used in place of μ in sections 8.2 and 8.3.

8.1.3 The thermodynamic functions

It remains to calculate the thermodynamic functions, U, C_V, P and so on. We shall do this explicitly in the degenerate limit only, and shall sketch graphs to indicate how the functions connect to the classical MB limit. Numerical methods are needed to compute the shape of such graphs.

Internal energy U. The internal energy can be evaluated from the distribution function $f(\varepsilon)$, simply using the direct expression

$$U = \sum_i n_i \varepsilon_i$$

In the present context this becomes

$$U = \int_0^\infty \varepsilon g(\varepsilon) f(\varepsilon) d\varepsilon \tag{8.7}$$

At $T = 0, f(\varepsilon)$ is the simple step function, so that the integral is readily evaluated as

$$U(0) = \int_0^\mu \varepsilon g(\varepsilon) d\varepsilon$$

Substituting $g(\varepsilon) = C\varepsilon^{1/2}$ for the fermion gas, this gives

$$\begin{aligned} U(0) &= \frac{2}{5} C\mu^{5/2} \\ &= \frac{3}{5} N\mu \end{aligned} \tag{8.8}$$

The result (8.8) follows without the need to remember C when it is recalled from the previous section that $N = \frac{2}{3}C\mu^{3/2}$ from the corresponding integral (8.4). This value of $U(0)$ represents a very large zero-point energy, an average of 0.6 μ per particle. It is a direct expression of the exclusion principle, that the particles are forced to occupy high energy states even at $T = 0$. (Fermions display tower-block mentality!)

Above $T = 0$, whilst yet in the degenerate limit $T \ll T_F$, the integral (8.7) can be evaluated using the method outlined in Appendix C. The result is

$$\begin{aligned} U &= U(0) + U(\text{th}) \\ &= \frac{3}{5}N\mu + (\pi^2/6)(k_B T)^2 g(\mu) + \cdots \end{aligned} \tag{8.9}$$

As expected, U remains dominated by the zero-point term $U(0)$, the second thermal term $U(\text{th})$ being small. This is a reasonable result. In hand-waving terms it follows from the subtle blurring of $f(\varepsilon)$ with rising T. At $T = 0$ we have seen that all states with energies below μ are full, whereas all those above are empty. At a low temperature T, from (8.2) or Fig. 8.1, it is evident that f changes from 1 to 0 over an energy span of order $k_B T$ around the Fermi energy, the occupation numbers of states outside this span being unchanged. Hence only a number of order $g(\mu) \times k_B T$ of the fermions have their energies changed, and the change in energy of these is of order $k_B T$. Therefore one would expect $U(\text{th}) \sim (k_B T)^2 g(\mu)$, as in (8.9).

Equation (8.9) is shown with the explicit factor $g(\mu)$, since this displays the correct physics, as just explained. It is precisely this 'density of states at the Fermi level' which enters many of the thermodynamic and the transport properties of a fermion gas. And (8.9) continues to give the right answer even when we are not talking about an ideal gas in three dimensions, and the density of states is not of the form (8.3).

Nevertheless, it is also interesting to return to the standard ideal gas. Substituting into (8.9) the appropriate density of states, (8.3), together with the expression (8.5) for the Fermi energy, we obtain for the thermal internal energy

$$U(\text{th}) = (\pi^2/2)Nk_B T \times (k_B T/\mu) \tag{8.10}$$

This is a useful way of writing the ideal gas result. We see that the thermal energy (omitting a numerical factor) is essentially the MB result multiplied by the 'Fermi factor' $(k_B T/\mu)$. And this factor is (a) small and (b) temperature dependent.

Heat capacity C_V. The hard work is now done. The heat capacity C_V follows immediately, since

$$\begin{aligned} C_V &= \mathrm{d}U/\mathrm{d}T \\ &= \mathrm{d}U(\text{th})/\mathrm{d}T \\ &= (\pi^2/3)k_B^2 T g(\mu) \end{aligned} \tag{8.11}$$

in the degenerate limit, obtained by differentiating (8.9). Hence at low temperatures one has a linear, and small, heat capacity. For the ideal gas (as in equation (8.10)) its magnitude is again of order the classical value Nk_B multiplied by the Fermi factor $(k_B T/\mu)$. The restriction of thermal excitation enforced by the FD statistics ensures that the heat capacity is a small effect.

Pressure P. The pressure is also readily deduced from U. In fact it remains true that for a gas of massive particles $P = \frac{2}{3}U/V$ independent of statistics. Hence PV and U are proportional. This relation follows from basic statistical ideas

$P = -(\partial U/\partial V)_S$	basic thermodynamics
$= -(\partial U/\partial V)_{\{n_i\}\text{fixed}}$	S depends on $\{n_i\}$ only
$= -\sum n_i(\partial \varepsilon/\partial V)$	compare section 2.3.2 – only the energy levels depend on V
$= \frac{2}{3}\sum n_i \varepsilon_i V$	see below
$= \frac{2}{3}U/V$	as stated!

The volume dependence of the energy levels depends on the dispersion relation alone. For our gas of particles of mass M, we know that $\varepsilon(k) \propto k^2$. But k, a dimension in k-space, is proportional to the reciprocal of the box size a, i.e. $k \propto V^{-1/3}$. Hence $\varepsilon \propto V^{-2/3}$. Each energy level depends on V with this same power law, so that we can replace $\partial\varepsilon/\partial V$ by $(-\frac{2}{3})\varepsilon/V$.

Therefore P follows the energy density U/V. The Fermi gas therefore is seen to exert a very large zero-point pressure $P(0) = \frac{2}{3}U(0)/V$. This plays an important role in considerations of stability in white-dwarf stars, where the pressure (of order 10^{17} atm) inhibits further collapse under the influence of gravitational forces. In the case of metals, electrostatic binding forces are large and the electrons are contained in the metal by the so-called work function of the metal, an energy which must be greater than μ for the electrons not to leak out.

Entropy S. The temperature dependence of S is readily obtained from C_V. We obtain a linear dependence on T (just as for C) in the degenerate region, S tending to zero as it should at the absolute zero. This passes towards the ln T variation in the classical

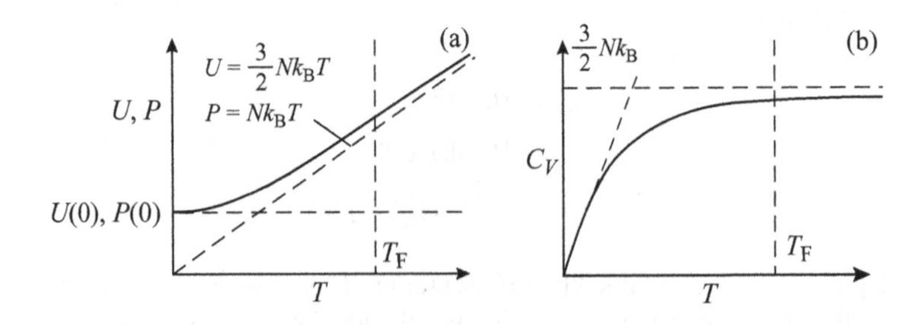

Fig. 8.3 The variation with temperature of the thermodynamic functions for an ideal FD gas. Note the low-temperature (quantum) and high-temperature (classical) limits. (a) Internal energy and pressure. (b) Heat capacity.

region. A full calculation of S or of F is fairly complicated, and will not be attempted here. (There is no partition function Z to help on the way). The best route is to use $S = k_B \ln \Omega$, together with the known form of $f(\varepsilon)$.

Figure 8.3 illustrates the schematic behaviour of the thermodynamic functions, U, P and C_V.

8.2 APPLICATION TO METALS

Since this is a major topic of any book on the physics of solids, this section will be brief. The free-electron model is surprisingly successful in describing both transport and equilibrium properties of the conduction electrons. The surprise is that the very large electrostatic interactions, both between pairs of electrons and also between an electron and the lattice ions, can for many purposes be neglected, or at any rate be treated as small. The outcome is that the conduction electrons can be modelled as an ideal FD gas, but the density of states is somewhat modified to take account of these interactions. See section 14.1 for a fuller discussion of these interactions.

Thus the heat capacity of a metal has a contribution from the conduction electrons which is precisely of the form (8.11). Since T_F is so large (typically 70 000 K, as we have seen) this term is very small at room temperature compared with Nk_B. And the lattice contribution to the heat capacity of a solid (Chapters 3 and 9) is itself of order $3Nk_B$. Therefore, as a consequence of FD statistics, the heat capacity of a metal is very similar to that of a non-metal – a result confirmed by experiment. At low temperatures, the lattice contribution is frozen out as T^3 in a crystalline solid (Chapter 9). Hence the linear electronic term becomes readily measurable at around 1 K. Its magnitude is well understood from equation (8.11). A transition metal has a large density of states (the d-electron states) at the Fermi energy compared to a 'good' metal like copper. So the electronic heat capacity of copper is much smaller than that of, say, platinum.

The contribution to the magnetic susceptibility of the conduction electrons is also strongly influenced by FD statistics. In section 3.1.4 we discussed the magnetization of a spin-$\frac{1}{2}$ *solid* obeying Boltzmann statistics. In weak fields, the result was

$$M = N\mu^2 B / k_B T \qquad \text{(8.12) and (3.10)}$$

(In this section we revert to the use of μ for magnetic moment; we shall use the alternative symbol ε_F for the Fermi energy.) When we come to consider the magnetic contribution from aligning the conduction electron spins in an applied field B, the situation is different. Most of the spins are unable to align because there are no available empty states.

The problem is illustrated in Fig. 8.4, relevant to $T = 0$. The only spins which realign in the field are those in the shaded states, numbering $\mu B \times \frac{1}{2} g(\varepsilon_F)$ the factor of $\frac{1}{2}$ arising since only half the states are in the spin-down band. Each of these electrons

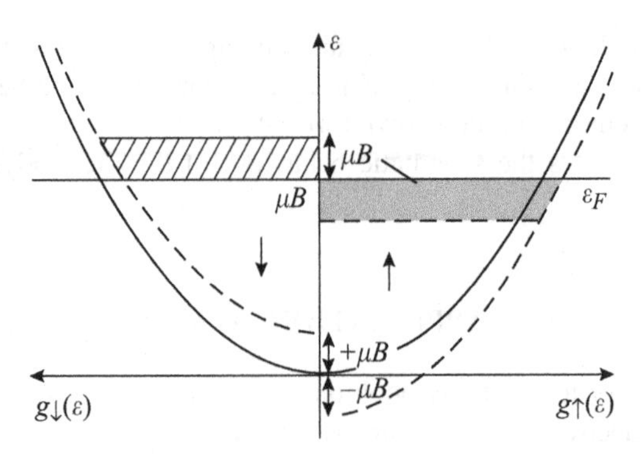

Fig. 8.4 Calculating the magnetization of a spin-$\frac{1}{2}$ FD gas at $T = 0$. The left-hand half of the figure gives the density of states for spin-down fermions, whereas the right-hand half is for spin-up fermions. The full curve shows the densities of states when $B = 0$, the dashed curves when a field B is applied. Only a small number of the spins close to the Fermi energy (indicated by the shaded region) are able to realign when B is applied.

has its spin reversed (i.e. changed by 2μ), giving a total magnetization of

$$M = \mu^2 B g(\varepsilon_F) \tag{8.13}$$

This expression is independent of T in the degenerate region, since the argument of Fig. 8.4 is unaffected by the slight blurring of the Fermi surface. Equation (8.13) is again a characteristic FD result. Since $g(\varepsilon_F) = 3N/2\varepsilon_F$ for the ideal gas, we see that the magnetization is reduced from the Boltzmann value, (8.12), by the usual Fermi factor $(k_B T/\varepsilon_F)$. Again the small magnitude and the lack of T-dependence of the magnetization are well confirmed by experiment, as is the dependence on $g(\varepsilon_F)$ in particular metals.

8.3 APPLICATION TO HELIUM-3

Helium-3 is a spin-$\frac{1}{2}$ atom, having a nuclear spin of $\frac{1}{2}$. Therefore ^{3}He gas is a fermion gas. However, as explained in Chapter 6, it never exists with a density high enough for FD statistics to be more than a small correction to the MB treatment. But that is not the end of the ^{3}He story, since it turns out that FD statistics can usefully be applied to *liquid* ^{3}He.

Again this idea should come as a surprise! There are very strong interactions between molecules in a liquid, and the atoms in liquid ^{3}He are no exception. One of the achievements of Landau was to develop the theory of a Fermi liquid, and to show that for many purposes it can be treated as a weakly interacting gas. However, the particles of the gas are not ^{3}He atoms, but are ^{3}He 'quasiparticles', entities which can be visualized as bare ^{3}He atoms dressed with the interactions of other nearby atoms. After all, if one atom in the liquid is moved, then a whole lot of other atoms must

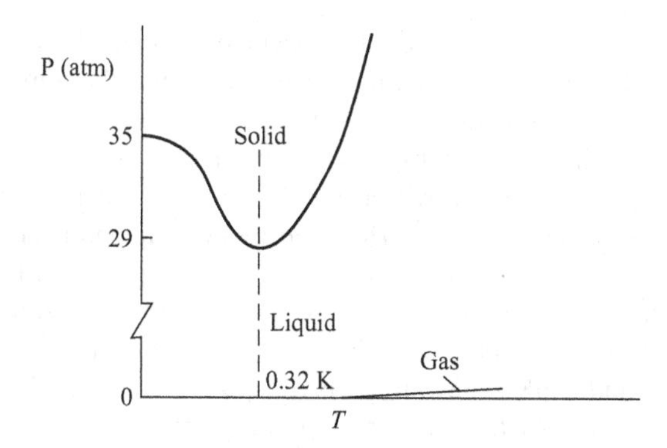

Fig. 8.5 The phase diagram for ^{3}He (not drawn to scale).

move in order for the liquid to re-establish uniform density. The Landau model ends up with a gas with the same number N of particles as atoms, but with the *effective mass* of each particle being larger than the bare mass. In liquid ^{3}He this effective mass is about 3–6 times the atomic mass of ^{3}He, depending on the applied pressure and hence on the density.

This simple picture of liquid ^{3}He is well supported by the measured thermodynamic properties. The heat capacity follows the general shape of Fig. 8.3, in particular being linear at low temperatures. But the value of T_F deduced from experiment requires, as suggested by Landau, a large effective mass. We give a fuller discussion of this topic in Chapter 14 (section 14.2). Similarly the nuclear spin magnetization is found to follow the form suggested in section 8.2 for spin magnetization of a fermion gas.

These ideas also help to explain the highly unusual phase diagram of ^{3}He, shown in Fig. 8.5. There are two points worthy of note, both relevant to statistical physics. The first is that ^{3}He remains liquid, at modest pressures, right to the absolute zero. In this the helium isotopes are unique. The explanation is that the binding energy of the solid is so weak that it can be overcome even by its zero-point vibrations, making the solid unstable until it is stiffened up by compression. This 'quantum manifestation' occurs in both ^{3}He and ^{4}He, showing that it has nothing to do with fermions or bosons. The second feature of interest occurs only in ^{3}He, namely the region of negative slope of the solid-liquid equilibrium line below 0.3 K. It is, therefore, specifically relevant to the fermion system.

An important result of thermodynamics is that such a slope can be related to entropy and volume changes using the *Clausius–Clapeyron equation* (see e.g. *Thermal Physics* by Finn, section 9.4). This states that the slope of the equilibrium line

$$\mathrm{d}P/\mathrm{d}T = \Delta S/\Delta V$$

where ΔS and ΔV are, respectively, the entropy and the volume changes which occur when the phase change takes place. Usually when a substance melts, its volume increases (ΔV positive) and its entropy increases (ΔS positive). Therefore $\mathrm{d}P/\mathrm{d}T$

is usually positive. A negative $\mathrm{d}P/\mathrm{d}T$ can occur in two situations. Occasionally it happens that the solid contracts on melting, ice to water being the commonest example. In the case of ^{3}He, however, the volume change is normal. The peculiarity of ^{3}He arises entirely from an unusual entropy variation. Below 0.3 K the solid entropy is bigger than the liquid, in other words the solid is more disordered than the liquid!

This arises because the entropy of ^{3}He at these low temperatures is due principally to its spin disorder. In zero magnetic field, the spins in the solid behave precisely as a spin-$\frac{1}{2}$ solid, as discussed in section 3.1. They are totally disordered, giving an entropy of $Nk_B \ln 2$ at all temperatures above a few mK. The spins order by their own interactions at around 2 mK. The entropy is illustrated schematically in Fig. 8.6. On the other hand the Fermi factor keeps the liquid entropy low. As in our discussion of the heat capacity, only a small number (of order $k_B T g(\varepsilon_F)$) of the spins are free to change their state. The exclusion principle ensures that the others are frozen in, and at $T = 0$ zero entropy is achieved by all states below the Fermi level being definitely full and those above it being definitely empty: hence Fig. 8.6. The liquid (FD gas) entropy is not of course limited to $Nk_B \ln 2$, so that at a high enough temperature, it crosses the solid curve and continues upward.

Finally we can note that this phenomenon is more than a theorist's re-creation. It forms the basis of 'Pomeranchuk cooling', a practical method for achieving temperatures as low as 2 mK. If liquid ^{3}He is precooled to a temperature somewhat below 0.3 K, and is then converted to solid by compression, refrigeration is produced. For instance if the conversion to solid were isothermal (T constant), Fig. 8.6 illustrates that a large amount of entropy would be extracted from the heat reservoir. On the other hand if the compression to form solid were adiabatic (S constant), the same S–T diagram shows that the temperature of the ^{3}He would reduce. With reference to the phase diagram (Fig. 8.5), the liquid–solid mixture would pass up the anomalous (negative slope) co-existence curve as the pressure increases, until all the helium has solidified. The entropy scale on Fig. 8.6 is quite large, so that this is an effective cooling method.

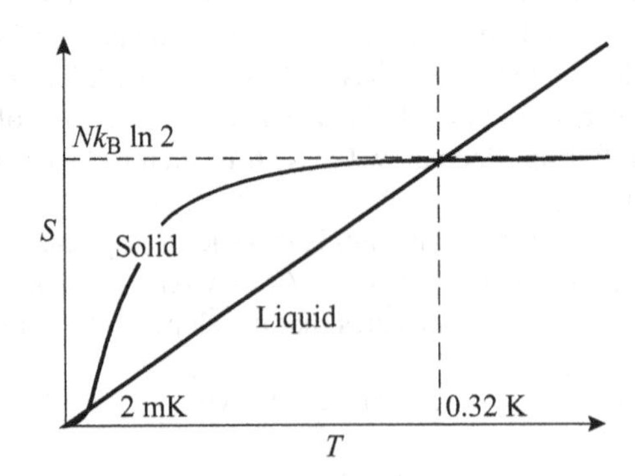

Fig. 8.6 Entropy–temperature curves for liquid and solid ^{3}He. In the anomalous (= unusual) region below 0.32 K the solid is more disordered than the liquid.

8.4 SUMMARY

This chapter discusses the properties of an ideal Fermi–Dirac gas.

1. Quantum statistics rather than MB statistics can be appropriate for gases at high density, low temperature and light particle mass.
2. The exclusion principle dominates FD statistics, since the maximum occupation number of a one-particle state is 1. States are full or empty.
3. At $T = 0$ there is a sharp Fermi energy, defined so that all lower energy states are full, and all higher energy states are empty.
4. This energy can be represented by a 'Fermi surface' in k-space
5. At higher temperatures, the Fermi surface becomes blurred, with states within $k_B T$ of the surface having intermediate occupation on average.
6. Many thermodynamic properties of the gas are related to those of the MB gas, reduced by a factor $k_B T/\varepsilon_F$.
7. Conduction electrons in metals provide a good example of FD properties, even though the density of states is not that of an ideal free gas.
8. Liquid helium-3 is (surprisingly) another candidate, for reasons which will be explored further in Chapter 14.
9. The phase diagram of helium-3 shows that the solid is more disordered than the liquid, a demonstration that the lack of flexibility in the FD gas inhibits spin disorder.

9

Bose–Einstein gases

This chapter discusses the properties of an ideal Bose–Einstein (BE) gas, which without any interactions nevertheless shows a remarkable phase transition, the 'Bose–Einstein condensation'. This property is relevant to liquid ^{4}He and to the behaviour of groups of 'cold atoms'. But another important application of BE statistics is to the 'phoney' boson gases, photons and phonons.

9.1 PROPERTIES OF AN IDEAL BOSE–EINSTEIN GAS

9.1.1 The Bose–Einstein distribution

In Chapter 5 ((5.11) and (5.13)) we derived the form of the distribution function, i.e. the number of particles per state of energy ε in thermal equilibrium. For a gas of ideal bosons the distribution is

$$f_{BE}(\varepsilon) = 1/[B\exp(\varepsilon/k_B T) - 1] \tag{9.1}$$

For clarity the subscript BE will be omitted in the rest of this chapter. The parameter B is to be determined from the number condition $\sum n_i = N$, which caused its appearance in the first place (see again Chapter 5). We shall discuss this number condition in terms of B for the boson gas, but it is entirely equivalent to the use of α or μ from the identities

$$B = \exp(-\alpha) = \exp(-\mu/k_B T)$$

Before coming to the determination of B, the main task of this section, we can observe some of its properties just from inspection of the distribution (9.1). To be specific, let us measure the one-particle energies ε using the ground state as the zero of energy, i.e. $\varepsilon_0 = 0$. Now to make physical sense, we know that for all ε the distribution $f(\varepsilon)$ must be positive. Because of the boson -1 in the denominator, this requires $B > 1$. If B were negative, then at least the ground-state occupation would be negative! Furthermore, if one were to suppose $B = 1$, then the ground-state occupation would be infinite, a possibility in a 'friendly' boson system since there is no exclusion principle. Hence we must have $B > 1$ to describe the finite number N of gas particles.

The value of B will be found to vary with density and temperature, as did μ in the discussion of FD statistics or α in the case of MB statistics. We can recognize from (9.1) that a simplification to the MB limit will occur when $B \gg 1$. Under these circumstances the -1 in the denominator will become insignificant for all values of ε, even for the ground state. Hence (9.1) could be replaced by the simple $f_{\text{MB}}(\varepsilon) = (1/B)\exp(-\varepsilon/k_{\text{B}}T)$.

The obvious approach to determining B, and the one used with success in the MB and FD cases, is to use the density of states approximation, and to replace the sum over all states by an integration over k or ε. So we enumerate the states by the usual function

$$g(k)\mathrm{d}k = V/(2\pi^3) \cdot 4\pi k^2 \mathrm{d}k \cdot G \qquad \text{(9.2) and (4.5)}$$

We shall for simplicity consider spin-0 bosons for which the spin factor $G = 1$. Making the substitution into (9.1) of $\varepsilon = \hbar^2k^2/2M$, as is appropriate for a gas of particles of mass M, we obtain

$$\begin{aligned} N &= \sum_i n_i \\ &= \sum_i g_i f_i \\ &= \int_0^\infty g(k)f(k)\mathrm{d}k \\ &= V/(2\pi)^3 \cdot 4\pi \int_0^\infty k^2\mathrm{d}k/[B\exp(\hbar^2k^2/2Mk_{\text{B}}T) - 1] \end{aligned}$$

or, after a little work,

$$N = Z \cdot F(B) \tag{9.3}$$

In (9.3) the factor Z is the same as the MB partition function for the gas (hence the notation), i.e. $Z = V(2\pi Mk_{\text{B}}T/h^2)^{3/2}$. The function $F(B)$ is defined by what is left in the equation, which by the substitution $y^2 = \hbar^2k^2/2Mk_{\text{B}}T$ is seen to be

$$F(B) = (4/\sqrt{\pi}) \int_0^\infty y^2\mathrm{d}y/[B\exp(y^2) - 1] \tag{9.4}$$

Equation (9.3) can equally well be written as $F(B) = A$, where A is the usual degeneracy parameter (defined as N/Z). A is given by the (N, V, T) macrostate conditions, so that B can be determined from a table of values of $F(B)$ against B, a numerical task since (9.4) cannot be readily inverted. So is the problem solved?

To take the good news first, this solution certainly makes sense at high T or at low density N/V, i.e. in the MB limit $A \ll 1$. In these circumstances, B is large, the -1 in the denominator of equation (9.4) is negligible, and the function $F(B)$ reduces simply to $1/B$. (This may be checked using the I_2 integral of Appendix C.) This reproduces the correct MB normalization, $1/B = N/Z$ (Chapter 6). Furthermore, as A rises towards unity, nothing obvious goes wrong, and the above treatment gives plausible and calculable deviations from MB behaviour. However, as the gas becomes highly degenerate ($A > 1$), a nonsense appears, as we shall now demonstrate.

9.1.2 The Bose–Einstein condensation

If we lower the temperature far enough, then an apparent contradiction occurs. The difficulty is that, although we know that the minimum value possible for B is $B = 1$, the function $F(B)$ remains finite as $B \to 1$. In fact it takes the value $F(1) = 2.612\ldots$. Hence if $A > 2.612$, then equation (9.3) has no acceptable solution. This condition corresponds to $T < T_B$, where T_B, the Bose temperature, is defined by $A(T_B) = 2.612$, i.e. by

$$T_B = (h^2/2\pi M k_B)(N/2.612V)^{2/3} \tag{9.5}$$

So, although our treatment makes sense when $T > T_B$, nevertheless when $T < T_B$ we apparently are not able to accommodate enough particles. The integral (9.3) is less than N even if $B = 1$.

The source of this difficulty lies in the total reliance on the density of states approximation. In particular, the approximation gives zero weight to the ground state at $k = 0$, which is effectively removed by the k^2 factor inside the integrand of (9.3). That is entirely valid if $g(k)$ and $f(k)$ are slowly varying functions of k. But we have already seen that in the region of $B \approx 1$, this is anything but the case. In fact $f(0)$ diverges as $B \to 1$.

A simplified (but essentially correct) solution to the problem is as follows. We group the states into two classes: (i) a ground-state group, consisting of the lowest few states – it makes no real difference exactly how many are taken. Assume g_0 such states at $\varepsilon = 0$, (ii) the remaining states of energy $\varepsilon > 0$. Since $f(\varepsilon)$ will be a smooth function over all $\varepsilon \neq 0$, this group may be described by the density of states approximation. Suppose that in thermal equilibrium there are n_0 bosons in the ground-state group, and $N(\text{th})$ in the higher states. Then the number condition (9.3) is replaced by

$$N = n_0 + N(\text{th}) \tag{9.6}$$

in which the ground-state occupation is

$$n_0 = g_0/(B - 1) \tag{9.7}$$

and in which the higher state occupations continue to be given by the same integral as in (9.3) (the lower limit of the integral is in principle changed from 0 to a small finite number, but this has no effect since, as noted above, the integrand vanishes at $k = 0$).

Equation (9.6) has radically different solutions above and below T_B. Above this temperature, the second term (corresponding to the vast majority of available states) can contain all the particles. We have $B > 1$ and hence $n_0 \approx 0$ from equation (9.7). Equation (9.6) becomes $N = N(\text{th})$, with $N(\text{th})$ given precisely by (9.3). Nothing has changed from the previous section.

Below T_B, however, the first term in (9.6) comes into play. We have $B \approx 1$, so that the second term becomes $N(\text{th}) = Z \times 2.612 = N(T/T_B)^{3/2}$, using the definition (9.5). The first term is given by the difference between N and $N(\text{th})$, i.e.

$$n_0 = N[1 - (T/T_B)^{3/2}] \tag{9.8}$$

This accommodation can be made with the merest perceptible change in B, B being of a magnitude $1 + O(1/N)$ throughout the low temperature region. Graphs of n_0 and of B are given in Figs. 9.1 and 9.2.

The properties of the 'condensate', the ground-state particles, are rather interesting. At $T = 0$, as expected, we have $n_0 = N$ and all the particles are in the ground state. However, even at any non-zero temperature below T_B we have a significant fraction

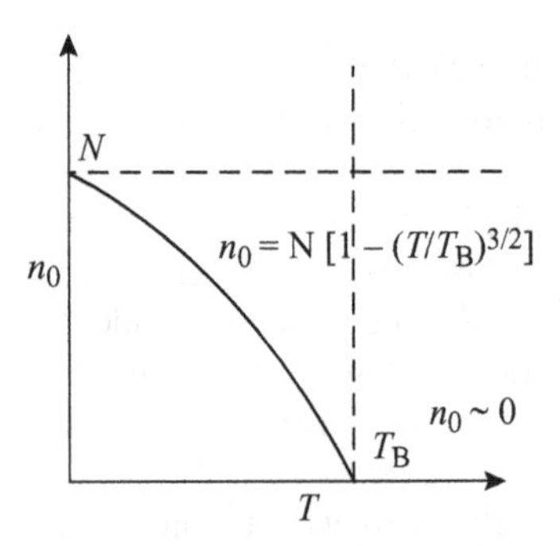

Fig. 9.1 The ground-state occupation n_0 for an ideal BE gas as a function of temperature. The ground state is not heavily occupied above the condensation temperature T_B, but below T_B it contains a significant fraction of the N bosons.

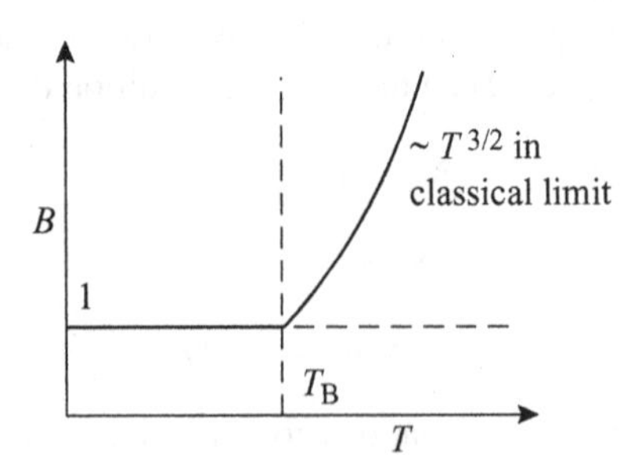

Fig. 9.2 The variation with T of the normalization parameter B for an ideal BE gas. At and below T_B, B becomes hooked up at a value (just greater than) unity.

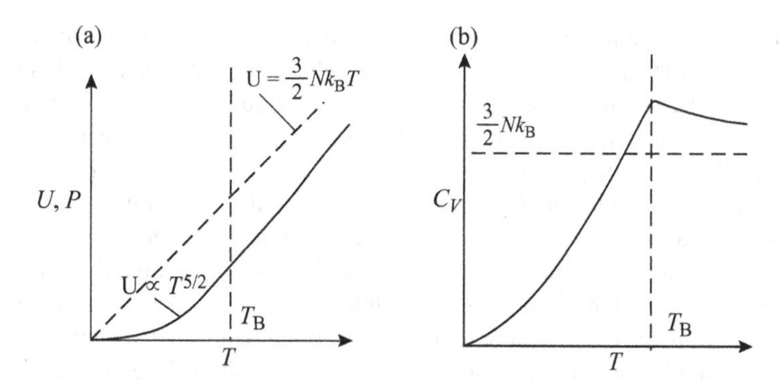

Fig. 9.3 The variation with temperature of the thermodynamic functions for an ideal BE gas. Compare Fig. 8.3 for the contrasting behaviour of an ideal FD gas. (a) Internal energy and pressure. (b) Heat capacity.

(given by (9.7)) already dumped into the ground state. And these particles do not contribute to the internal energy, heat capacity, entropy, etc. The condensate fraction is already in its absolute-zero state.

The thermodynamic functions obtained from the occupation of the states are illustrated in Fig. 9.3(a) and (b). The characteristic BE result for any function is:

1. A simple power-law behaviour at $T < T_B$, arising from the $T^{3/2}$ dependence of $N(\text{th})$ together with the (effective) constancy of B. The variation of U is $T^{5/2}$, and so correspondingly is the variation of P (since $P = 2U/3V$ as shown in section 8.1.3). The heat capacity C_V, varies at $T^{3/2}$ up to a maximum value of about 1.9 Nk_B at T_B.
2. A phase transition at T_B, albeit a gentle one. In common parlance it is only a third-order transition, since the discontinuity is merely in $\mathrm{d}C/\mathrm{d}T$, a third-order derivative of the free energy; the second-order C and the first-order S are continuous.
3. A gradual tendency to the classical result when $T \gg T_B$, from the opposite side from the FD gas (compare Fig. 8.3).

Viewed from the perspective of high temperatures, the onset at T_B of the filling of the ground state is a sudden (and perhaps unexpected) phenomenon, like all phase transitions. It is referred to as the Bose–Einstein condensation. In entropy terms, it is an ordering in k-space (i.e. in momentum space) rather than the usual type of phase transition to a solid, which is an ordering in real space.

9.2 APPLICATION TO HELIUM-4

The atom ^{4}He is a spin-0 boson. So just as we saw that FD statistics were relevant to ^{3}He, so we should expect BE statistics to apply to ^{4}He. Again, BE statistics are in the extreme only a small correction for ^{4}He gas, but their relevance is to the liquid.

Liquid ^{4}He remains fluid to the lowest temperatures at pressures below about 25 atm, for reasons discussed in section 8.3. Its phase diagram is sketched in Fig. 9.4. The outstanding feature is the existence of a liquid–liquid phase transition at around 2 K from an ordinary liquid (HeI) at high temperature to an extraordinary 'superfluid' phase (HeII) at low temperatures. The transition is evidenced by a heat capacity anomaly at the transition temperature, commonly called the lambda-point because of the resemblance of the $C-T$ curve to the Greek letter λ. The curve is shown in Fig. 9.5, with the corresponding $S-T$ curve in Fig. 9.6. The nature of the singularity is 'logarithmic', in that although C_P is infinite, its integral across the transition is finite so there is no latent heat associated with the transition. The entropy curve is continuous, but it does have a vertical tangent at the transition. The explanation of the very flat solid–liquid line on the phase diagram can again (as in ^{3}He) be understood from the Clausius–Clapeyron equation. Below the lambda-point, both liquid (because of superfluid ordering) and solid (^{4}He has zero spin, unlike ^{3}He) are highly ordered and have virtually zero entropy. Hence the entropy difference between the phases is almost zero so that the phase equilibrium line has almost zero slope.

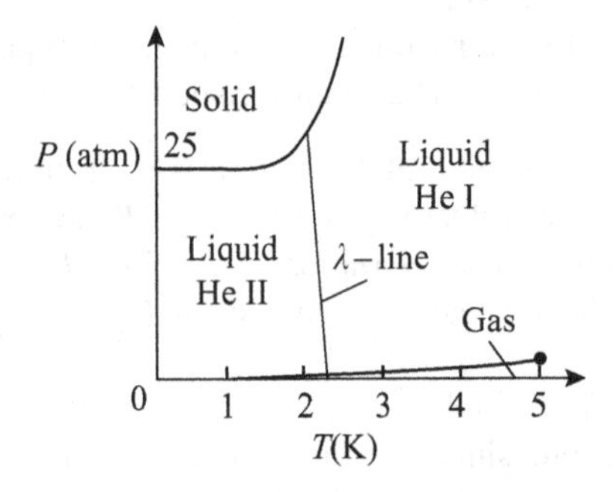

Fig. 9.4 The phase diagram of ^{4}He.

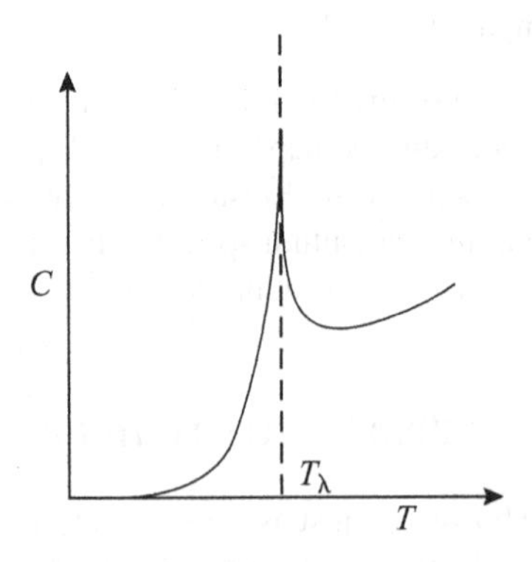

Fig. 9.5 The heat capacity of liquid ^{4}He as a function of T, showing the lambda anomaly.

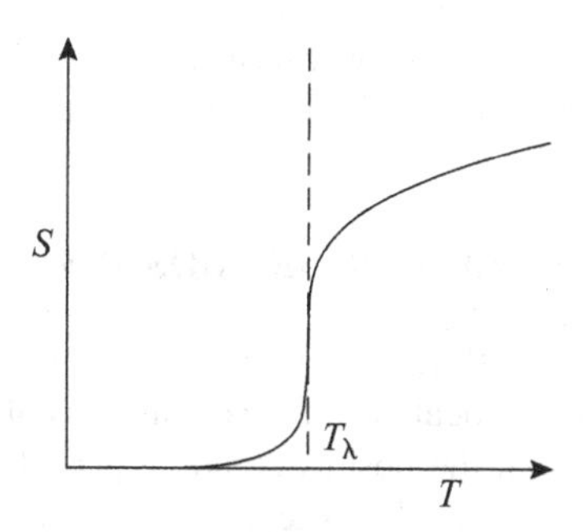

Fig. 9.6 The variation of entropy with temperature for liquid ^{4}He, showing a vertical tangent at T_λ.

All this bears some relation to the BE gas, although the actual transition in ^{4}He is more sudden than that of the ideal gas. It is believed that the difference can be attributed to the strong interactions between ^{4}He atoms in the liquid (see Chapter 14, section 14.3, for a fuller discussion). Furthermore, the transition takes place at a temperature of the correct order of magnitude, the value of T_B for an ideal gas of bosons with the bare ^{4}He mass and the density of liquid ^{4}He being about 3.1 K.

Even more reminiscent of the BE gas is the nature of the superfluid HeII. Its properties are well described by a 'two-fluid model', in which the liquid is thought of as a mixture of two interpenetrating fluids. One of these, the normal fluid, has the properties of a conventional fluid, very like the thermal N(th) fraction of the BE gas, (9.6). Whereas the other fluid, the superfluid, behaves like the condensate fraction n_0, in that it carries no entropy and displays no friction.

An excellent example of the two-fluid properties arises when the viscosity of liquid HeII is measured. The answer is found to depend dramatically on the type of experimental method.

1. If viscosity is measured by the drag on a rotating disc or a vibrating wire, then a reasonable value is found, similar to that for HeI, although reducing somewhat as T is lowered well below T_λ.
2. If, however, the measurement is made by allowing the helium to flow through tubes, one finds the astonishing answer zero for the viscosity. The fluid can flow with alarming ease (up to some critical velocity) through the thinnest channels; hence the term 'superfluid'. It can even syphon itself over the edge of a container through superflow in the surface film.

The explanation in terms of the two fluids is clear. The drag methods will pick out the most viscous fluid, and that is the normal fluid. The gradual reduction below T_λ arises from the gradual reduction of the normal fluid concentration (cf. N(th)) as T is lowered. On the other hand the flow experiments pick out the smallest viscosity. Only the superfluid flows through a fine tube, and this has zero viscosity since it carries no entropy. A demonstration that it is the (effectively $T = 0$) superfluid only which flows is to observe that the flow produces a temperature difference between the ends of the flow tube. The source reservoir warms up as it is depleted of superfluid whereas the receiving reservoir cools down as the superfluid enters.

9.3 PHONEY BOSONS

A boson is a particle of spin $0, 1, 2, \ldots$ by definition. Our discussion so far in the chapter has been based on an ideal gas of 'real' particles of mass M. However, there is another case, which is that the bosons could be particles with no rest mass. The obvious example is a photon gas. An evacuated box is never 'empty', in that it will contain electromagnetic radiation in thermal equilibrium with the walls of the box. Such thermal equilibrium radiation is called black-body radiation, and it may best be considered as a BE gas, an ideal gas of photons. The second analogous example is that the lattice vibrations of a solid can be modelled similarly as a gas of sound waves, or 'phonons', in a box which is the boundary of the solid.

The new feature of both of these gases is that the particles of the gas are massless. This has the importance that the *number* of the gas particles is not fixed. The particles may be, and are, destroyed and created continually. There is no conservation of N in the macrostate and the system is defined not by (N, V, T) but simply by (V, T). The empty box will come to equilibrium with a certain average number and energy of photons (or phonons) which are dependent only on the temperature and the volume of the box. So when we derive the thermal equilibrium distribution for the gas, there is a change. The usual number restriction $\sum n_i = N$, (5.7), does not enter, and the derivation of the distribution (section 5.4.2) is to be followed through without it. The answer is almost self-evident.

Since there is no N restriction, there is no α (so also no B, and no chemical potential μ). But the rest of the derivation is as before, and we obtain the 'modified BE' distribution

$$f(\varepsilon) = 1/[\exp(-\beta\varepsilon) - 1] \tag{9.9}$$

with as usual $\beta = -1/k_{\mathrm{B}}T$. We shall now apply this result to photons and phonons.

9.3.1 Photons and black-body radiation

This is an important topic. The radiation may be characterized by its 'spectral density', $u(\nu)$, defined such that $u(\nu)\delta\nu$ is the energy of the radiation with frequencies between ν and $\nu + \delta\nu$. This spectral density was studied experimentally and theoretically in the last century and was the spur for Planck in the early 1900s to first postulate quantization, an idea then taken up by Einstein. The Planck radiation law, of great historical and practical importance, can be readily derived using the ideas of this chapter, as we shall now see.

The modified BE distribution (9.9) tells us the number of photons per state of energy ε in thermal equilibrium at temperature T. In order to calculate the spectral density, we also need to know, (i) how many states there are in the frequency range of interest, and (ii) how the frequency ν of a photon relates to its energy ε. The second question is immediately answered by $\varepsilon = h\nu$. The first is yet another straightforward example of fitting waves into boxes.

The density of photon states in k is given by (4.5), with the polarization factor $G = 2$. Photons, since they are spin-1, massless bosons, have two polarization states; in classical terms electromagnetic waves are transverse, giving left- or right-hand polarizations, but there is no longitudinal wave. Hence

$$g(k)\delta k = V/(2\pi)^3 \cdot 4\pi k^2 \delta k \cdot 2 \tag{9.10}$$

We wish to translate (9.10) to a density of states in frequency ν, corresponding to the required spectral energy density. This is readily and accurately achieved for photons in vacuum, since $\nu = ck/2\pi$, where c is the speed of light. Making the change of variables, (9.10) becomes

$$g(\nu)\delta\nu = V \cdot 8\pi\nu^2 \delta\nu / c^3 \tag{9.11}$$

The answer now follows at once. The energy in a range is the number of states in that range $\times$ the number of photons per state $\times$ the energy per photon. That is

$$\begin{aligned} u(\nu)\delta\nu &= g(\nu)\delta\nu \times f(\nu) \times \varepsilon(\nu) \\ &= V \cdot 8\pi\nu^2\delta\nu/c^3 \times 1/[\exp(h\nu/k_B T) - 1] \times h\nu \\ &= V \cdot \frac{8\pi h\nu^3 \delta\nu}{c^3} \cdot \frac{1}{[\exp(h\nu/k_B T) - 1]} \end{aligned} \tag{9.12}$$

Equation (9.12) is the celebrated Planck radiation formula. It is drawn in Fig. 9.7 for three different temperatures. We can make several comments.

1. This is not how Planck derived it! Photons had not been invented in 1900. His argument was based on a localized oscillator model, in which each of the $g(\nu)\delta\nu$ oscillator modes had an average thermal energy *not* of $k_B T$, the classical incorrect result, but of $h\nu/[\exp(h\nu/k_B T) - 1]$ as derived in Chapter 3 (essentially (3.13) ignoring zero-point energy). The modern derivation is much to be preferred.
2. The Planck law is in excellent agreement with experiment. One of the features concerns the maximum in $u(\nu)$. Experiment (effectively Wien's law) shows that $\nu_{\max}$ is proportional to T. This is evident from (9.12), since the maximum will occur at a fixed value of the dimensionless variable $y = h\nu/k_B T$. In fact $y_{\max}$ is a little less than 3 (see Exercise 9.5).
3. Another experimental property is that the total energy in the radiation is proportional to T^4. This T^4 law follows from an integration of the Planck law (9.12) over

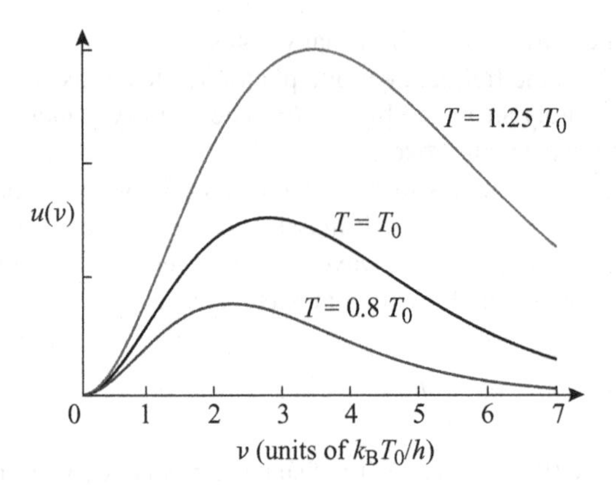

Fig. 9.7 The spectral distribution of energy in black-body radiation. The Planck formula (9.12) for the energy density as a function of frequency is shown at three different temperatures.

all possible frequencies. We have for the (internal) energy per unit volume

$$\begin{aligned} U/V &= (8\pi h/c^3) \int_0^\infty \nu^3 \mathrm{d}\nu/[\exp(h\nu/k_\mathrm{B}T) - 1] \\ &= (8\pi h/c^3)(k_\mathrm{B}T/h)^4 \int_0^\infty y^3 \mathrm{d}y/[\exp(y) - 1] \end{aligned} \tag{9.13}$$

The definite integral in (9.13) has the ('well-known') value of $\pi^4/15$. The energy U is represented by the areas under the curves in Fig. 9.7, which display the rapid variation of U and T.

4. Once U/V is known, two other properties follow. One is the energy flux radiated from a black body, a very accessible experimental quantity. It is defined as the energy per second leaving a small hole of unit area in the wall of the box, assuming no inward flux. (If the temperatures of the box and of its surroundings are the same, then of course there will be no *net* flux – usually a difference between flux in and flux out is the directly measurable quantity.) Since all the photons are moving with the same speed c, the number crossing unit area in unit time is $\frac{1}{4}(N/V)c$, where N/V is the (average) number density of photons. The factor $\frac{1}{4}$ comes from the appropriate angular integration, as in the corresponding standard problem in gas kinetic theory. Hence the energy crossing unit area in unit time is $\frac{1}{4}(U/V)c \equiv \sigma T^4$. This result is the Stefan–Boltzmann law, and the value of σ deduced from (9.13) is in excellent agreement with experiment.
5. The second property is the pressure, the 'radiation pressure'. We have already seen (section 8.1.3) that for a gas of massive particles, $PV = 2U/3$, this relationship following from the dispersion relation $\varepsilon \propto k^2 \propto V^{-2/3}$. For the photon gas, the dispersion relation is different, namely $\varepsilon = h\nu = ck/2\pi$. Hence $\varepsilon \propto k \propto V^{-1/3}$

in this case, and therefore (following the argument of section 8.1.3) $P = U/3V$. The radiation pressure is simply one-third of the energy per unit volume.

6. Finally let us consider the range of validity of the T^4 law. No high-temperature limit is evident, since photon states exist well above the highest conceivable thermally excited frequencies. No problem exists here (unlike the phonon case discussed below). At low temperatures, however, we should see the breakdown of the density of states approximation. The T^4 law is based on the replacement of a sum over photon states by an integral over a smooth density of states. This will clearly become invalid in a box of linear dimension a when $k_B T < \Delta\varepsilon$, the energy level spacing, i.e. for $k_B T < ch/a$. This corresponds roughly to $T < 1$ K for a box of side 1 cm, but $T = 1$ K is just too low a temperature for radiation to be important in practice anyway.

9.3.2 Phonons and lattice vibrations

The same ideas may be applied to a discussion of the thermal lattice vibrations of a simple atomic solid. We have already noted in Chapter 3 the inappropriateness of the localized oscillator model, since the motion of one atom in the solid is strongly coupled to that of another. The weakly coupled vibrational modes are sound waves, and hence it is appealing to consider the vibrations of the lump of solid as the motion of a gas of sound waves in a box. Hence the similarity to the previous section, where a box of electromagnetic waves was discussed.

The weakly interacting gas particles, quantized sound waves, are called phonons. And the total internal energy U of the phonon gas can be calculated in an analogous way to equation (9.13). However, there are three important differences in comparison with the photon gas:

1. Sound waves have $G = 3$, not $G = 2$. The three polarizations arise from three-dimensional motion of the atoms in the solid. In the long wavelength limit, there are two transverse polarization modes and one longitudinal mode. Hence (9.13) should be multiplied by a factor 3/2.
2. The dispersion relation is altered. To a first approximation we can simply use $\varepsilon = h\nu = c_S k/2\pi$, where c_S is a suitable average velocity of sound. This will simply replace c by c_S in (9.13). In practice, as any book on solid-state physics will reveal, this is a passable approximation for long wavelength (low k) phonons, although even there the velocities of transverse and longitudinal modes are very different. However, the linear dispersion must fundamentally fail at large values of k, a point related to the following.
3. There are only a limited number of phonon modes in the solid. The solid is not an elastic continuum, but consists of N atoms, so that the wavefunction of a phonon only has physical meaning (the atomic displacement) at these N atomic sites. Hence there are only $3N$ possible modes. Compare the localized oscillator model, which similarly considered $3N$ oscillators. As indicated above, the correct phonon dispersion relation will fully take care of this problem. However, this means going

back to (4.5) and using a computer to fold in the complicated dispersion relation. An approximate fudge (the Debye model) is to maintain the linear dispersion relation, but to cut off the density of states above a Debye cut-off frequency ν_D, defined so that

$$3N = \int_0^{\nu_D} g(\nu)\mathrm{d}\nu$$

Although not in detail correct, the Debye model contains most of the important physics, and it gives the correct heat capacity of typical solids to, say, 10%. The result for U is a modified version of (9.13)

$$U = V(12\pi h/c_S^3)(k_B T/h)^4 \int_0^{y(T)} y^3\mathrm{d}y/[\exp(y) - 1] \tag{9.14}$$

where $y(T) = h\nu_D/k_B T$. The heat capacity C, shown in Fig. 9.8, is obtained by differentiating (9.14) with respect to T.

The introduction of the cut-off gives a scale temperature θ to the problem, defined by $k_B\theta = h\nu_D$. For most common solids θ has a value around room temperature (lower for lead, higher for diamond!). At high temperatures, $T \gg \theta$ or $y(T) \ll 1$, the cut-off ensures that the classical limit is recovered, yielding $U = 3Nk_B T$ and $C = 3Nk_B$ (see Exercise 9.6). On the other hand, at low temperatures, $T \ll \theta$ or $y(T) \to \infty$, we recover the result analogous to black-body radiation. The integral in (9.14) again becomes a constant, and we have $U \propto T^4$ and correspondingly $C \propto T^3$. This is the so-called Debye T^3 law, and is in good agreement with the experimental heat capacity of a crystalline solid at low temperatures.

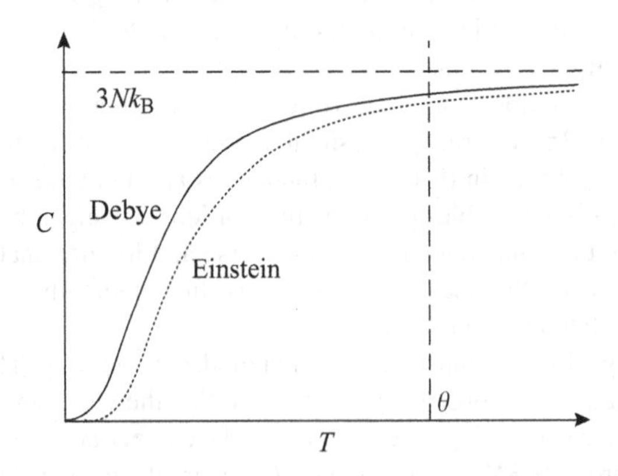

Fig. 9.8 The variation with T of the lattice heat capacity of a solid in the Debye approximation. Full curve, the Debye model (from (9.14)). Dotted curve, the Einstein model (Fig. 3.10), for comparison.

9.4 A NOTE ABOUT COLD ATOMS

We have seen that quantum statistics are needed at high densities and low temperatures. In recent years, scientists have developed some entirely new methods for cooling comparatively small (albeit macroscopic) numbers of gaseous atoms, usually alkali metal atoms such as sodium or rubidium.

Space (and possibly the author's expertise?) precludes a detailed discussion of this exciting field, exciting enough to generate the 1998 Nobel prize for three of its inventors. The techniques involve the clever use of laser beams to slow down a group of the atoms. Typically six beams are directed to the group from all sides, operating at a special tuned frequency such that atoms moving towards the beam are slowed, but those moving towards it are not; this trick (charmingly called 'optical molasses') depends on the energy level structure of the atoms and exploits the Doppler shift of the moving atoms. Using such techniques, typically billions of atoms are slowed down so that their translational kinetic energy corresponds to the microkelvin regime.

But the story doesn't end there. The atoms can also be confined within a limited volume using a magnetic field trap, a zero field 'bottle', exploiting the magnetic moment of the atoms. Finally, the hotter atoms in the trap can be 'evaporated' by lowering the edges of this trap, leaving an array of even colder atoms (corresponding to several nanokelvin). And these atoms can be observed optically or by other means.

The bottom line is that under these unusual conditions, quantum coherence and, in the case of bosonic atoms, Bose–Einstein condensation is observed. The simple theory of this chapter fits the observed facts very well, since these groups of cold atoms are sufficiently dilute that they are indeed examples of the weakly-interacting particles on which our statistics is based.

9.5 SUMMARY

This chapter discusses the properties of an ideal Bose–Einstein gas together with applications to systems of interest.

1. Quantum statistics rather than MB statistics are appropriate for gases at high density and low temperature.
2. The friendly nature of bosons leads to a Bose–Einstein condensation under these conditions. The condensation implies coherent behaviour of all the particles involved.
3. The ideal BE gas below T_B can be visualised on a two-fluid model of superfluid and normal fluid.
4. This picture is useful in considering the properties of liquid ^{4}He, which shows a superfluid transition.
5. Interactions between atoms are strong in the liquid, so that the ideal gas model does not apply in detail.
6. The ideal gas model does apply well to assemblies of cold bosonic atoms.

7. Modified BE statistics describes the properties of assemblies of massless bosons, for which there is no conservation of particle number.
8. This form of statistics gives an excellent understanding of 'black-body radiation' as photons in thermal equilibrium.
9. A similar treatment of sound waves (as weakly-interacting phonons) enables us to understand and compute the thermal properties of crystalline solids.

10

Entropy in other situations

As the chart in the front of the book shows, we have now completed our elementary study of the thermal equilibrium properties of ideal solids and gases. However, it would be a pity to stop here, since statistical physics has plenty more to say about other types of system also. In this chapter we shall look again at entropy, and shall discuss the statistics of a system in which the macrostate specifies T rather than U. This generalization will help us to discuss vacancies in solids in this chapter, and phase transitions in the next.

10.1 ENTROPY AND DISORDER

In Chapter 1, we took as a statistical definition of entropy the relation $S = k_B \ln \Omega$ (equation (1.5)). Since many verifiable results have followed, we may by now have much confidence in the approach. In this chapter we study some further consequences of the relation.

10.1.1 Isotopic disorder

One simple form of disorder in a solid is isotopic disorder. For instance a block of copper consists of a mixture of ^{63}Cu and ^{65}Cu isotopes. Therefore, if the isotopes are randomly distributed on the lattice sites, there will be a large amount of disorder associated with all the possible arrangements of the isotopes.

Consider a solid whose N atoms have a proportion P_L of isotope L. In other words, there are $N_L = P_L N$ atoms of isotope L, and $\sum_L N_L = N$, where the sum goes over all the isotopes. The number of arrangements of isotopes on the N sites is given by the well-trodden third problem of Appendix A. It is

$$\Omega = N!/\prod_L N_L! \tag{10.1}$$

The calculation of S for this situation goes as follows:

$$\begin{aligned}
S &= k_B \ln \Omega && \text{definition (1.5)} \\
&= k_B \left(\ln N! - \sum_L \ln N_L! \right) && \text{from (10.1)} \\
&= k_B \left(N \ln N - \sum_L N_L \ln N_L \right) && \text{Stirling's approx., also } \sum N_L = N \\
&= -k_B \left[\sum_L N_L (\ln N_L - \ln N) \right] && \text{putting } N = \sum N_L \\
&= -N k_B \sum_L P_L \ln P_L && \qquad (10.2)
\end{aligned}$$

The answer (10.2) is a nice simple one. For instance if we were to have a 50–50 mixture of two isotopes, it would give an 'entropy of mixing' of $S = Nk_B \ln 2$, a not unexpected result, analogous to tossing pennies (Appendix B).

Whether this entropy is observable is another matter. In fact our block of copper does not separate out under gravity to have all the ^{65}Cu at the bottom and all the ^{63}Cu at the top, however cold it is made! Rather this disorder is frozen in, and a cold piece of copper is in a metastable state which contains this fixed amount of disorder. Therefore the entropy of isotopic disorder is usually omitted from consideration, since it has no influence on the thermal properties. Metastable states of this sort do not violate the third law of thermodynamics, since no entropy *changes* occur near the absolute zero.

Actually partial isotopic separation does occur in just one case, that of liquid helium. A liquid mixture of ^{3}He and ^{4}He is entirely random above 0.8 K. However, when it is cooled below this temperature, a phase separation occurs to give a solution of almost pure ^{3}He floating on top of a dilute solution of about 6% ^{3}He in ^{4}He. This phase separation is important to low-temperature physicists, not least because the difference in thermal properties of the ^{3}He between the two phases forms the basis of a 'dilution refrigerator', the work-horse cooling method for reaching 5 mK nowadays.

10.1.2 Localized particles

A similar calculation can be made for the entropy of an assembly of N localized particles in thermal equilibrium, the topic of Chapter 2. We have seen (equation (2.4)) that Ω, the total number of microstates, can be replaced by t^*, the number of microstates associated with the most probable distribution. Hence the entropy is given by $S = k_B \ln t^*$, with $t^* = N!/\Pi n_j^*$ (equation (2.3)). The evaluation of S

exactly parallels the derivation of (10.2), the result being

$$S = -Nk_B \sum_j P_j \ln P_j \tag{10.3}$$

where $P_j = n_j/N$ (strictly n_j^*/N) is defined to be the fraction of the N particles in the state j. This important equation can be taken as an alternative statistical definition of entropy, and indeed it can be applied with greater generality than (1.5).

10.1.3 Gases

The same approach, of expressing S directly in terms of the distribution, can be adopted for gases, using the expressions for t^* derived in Chapter 5. Again using Stirling's approximation together with a little rearrangement, one obtains the following results:

For FD gases, from (5.4),

$$S = k_B \sum_i g_i[-f_i \ln f_i - (1 - f_i)\ln(1 - f_i)] \tag{10.4}$$

For BE gases, from (5.5),

$$S = k_B \sum_i g_i[-f_i \ln f_i + (1 + f_i)\ln(1 + f_i)] \tag{10.5}$$

For MB gases, from (5.6), or from the dilute ($f_i \ll 1$) limit of (10.4) or (10.5),

$$S = k_B \sum_i g_i[-f_i \ln f_i] + Nk_B \tag{10.6}$$

In these equations, the sums are over all groups i of states; however, another way of writing $\sum_i g_i$ is simply as a sum over all states. Therefore there is much similarity between the first (common) term of these expressions and (10.3) for localized particles. We may write (10.3) as

$$S = k_B \sum_j [-n_j \ln n_j] + Nk_B \ln N$$

Recognizing that n_j and f_i are both defined as the number of particles per state (the filling factor), the similarity is explicit. In fact the only difference between equation (10.3) for localized particles and (10.6) for MB gas particles is an addition of $k_B \ln N!$ to the entropy, as we should expect from the different extensivity of the two cases (section 6.3.3).

The second terms in the FD and BE expressions are intriguing. Equation (10.4) displays that in an FD gas, one not only considers the disorder of the particles filling the states (the first term), but also of the empty states (the second term). Both terms vanish if $f_i = 0$ or 1, as is the case for each state at $T = 0$. Intuition concerning (10.5) is not so clearcut, except to note that some such term is a clear necessity to ensure that $S > 0$ when we now allow f_i to be >1.

10.2 AN ASSEMBLY AT FIXED TEMPERATURE

The statistical method adopted earlier in the book has been based on an assembly of particles in a macrostate of fixed number N, volume V and internal energy U. Of course for a large assembly, the specification of any macrostate in practice determines all the other thermodynamic quantities, such as temperature T, pressure P and chemical potential μ. So we have cheerfully applied our method to assemblies with a macrostate specified by (N, V, T), with little fear of anything going wrong. However, when we look back at the statistical method and its relation to entropy in particular, we shall be pointed towards a more general statistical approach, one that can be applied to new types of assembly.

Our old method is based on the properties of an *isolated system* (mechanically isolated since V is constant, thermally isolated since in addition U is constant, and of fixed particle number N). One thing we really know about an isolated system is that any internal rearrangements it makes, any spontaneous changes, will always be such as to *increase the entropy* of the system. That is the content of the second law of thermodynamics. The entropy of the equilibrium state of the isolated system is a maximum.

Now our statistical method can be viewed in exactly this way, as a theorist making changes until he finds the state of maximum entropy! We have seen that the equilibrium distribution is that with the largest number of microstates, i.e. with the maximum $t(\{n_j\})$. Our statistical mechanician fiddles with the distribution numbers until he finds the maximum $t(= t^*)$. But since in the equilibrium state $S = k_B \ln t^*$, this is just the same as the maximum entropy idea of the second law. The example of section 1.6.2 is of direct relevance here. A system prepared with a non-equilibrium distribution, (i) will increase its t by adjusting the distribution, and (ii) will increase its entropy.

Having reminded ourselves of how the thermodynamic view (together with $S = k_B \ln \Omega$) is related to the statistical method for an isolated system, we now turn to consider a system at fixed temperature. Consider an assembly described by an (N, V, T) macrostate.

The condition for thermodynamic equilibrium of the (N, V, T) system is equally secure. *The free energy F must be a minimum in equilibrium.* This condition (e.g. *Thermal Physics*, by Finn, section 6.4) is again a statement of the second law. The entropy of the *universe* (i.e. of the system together with a large heat bath at temperature T) must not decrease when heat flows in or out of the system to maintain its temperature. This can be readily shown to imply that the free energy of the *system* must not increase, i.e. that it is a minimum in equilibrium.

This then suggests another approach to the statistical physics for an assembly in a way (N, V, T) macrostate. We should make (allowable) adjustments in the distribution numbers $\{n_j\}$ until F is a minimum; and this minimum in F will describe the thermodynamic state, with the corresponding $\{n_j^*\}$ being the equilibrium distribution. In practice, this method is straightforward for the sort of systems we have discussed previously, since we can calculate F as a function of the distribution from

$$F = U - TS = U(\{n_j\}) - k_B T \ln t(\{n_j\}) \tag{10.7}$$

An example follows in the next section. But in addition the new method enables some new problems to be attacked.

10.2.1 Distinguishable particles revisited

The new statistical method based on minimizing F (equation (10.7)) may be illustrated by re-deriving the Boltzmann distribution for the situation of Chapter 2. We have N weakly interacting localized particles at a temperature T. The states of one particle are as before labelled by j.

The free energy F is given as in (10.7) by

$$\begin{aligned} F &= U - TS \\ &= \sum_j n_j \varepsilon_j - k_B T \ln t(\{n_j\}) \end{aligned} \tag{10.8}$$

We require to minimize F in this equation, subject now to only *one* condition, namely that $\sum n_j = N$. (There is no energy condition now since U is not restricted.) The mathematical steps are similar to those in section 2.1. The number of microstates $t(\{n_j\})$, given by (2.3), is substituted into (10.8). This expression for F is then differentiated, using Stirling's approximation as in (2.6), to give

$$\mathrm{d}F = \sum_j [\varepsilon_j + k_B T \ln n_j]\, \mathrm{d}n_j \tag{10.9}$$

Using the Lagrange method to find the minimum, we set $\mathrm{d}F - \alpha \mathrm{d}N = 0$, to take account of the number condition $\sum n_j = N$. (We have chosen to write the multiplier as $-\alpha$ simply in order to ensure below that α has its usual physical identification!) Substituting $\mathrm{d}F$ from (10.9) and removing the summation sign, the whole point of Lagrange's method, we obtain for the equilibrium distribution

$$\varepsilon_j + k_B T \ln n_j^* - \alpha = 0$$

i.e.

$$n_j^* = \exp(\alpha - \varepsilon_j / k_B T) \tag{10.10}$$

As we should expect, this is again the Boltzmann distribution. However, there is an important difference from (2.12), in that now there is no β to be discussed. The $-1/k_BT$ comes in from the outset, from the macrostate. The multiplier α has the same meaning as earlier, and is determined by substituting back into the number condition $\sum n_j = N$.

10.3 VACANCIES IN SOLIDS

We have just seen that the free energy method gives the correct answer to an already familiar problem. That is comforting, but not very exciting. Of greater importance is that we can now tackle new problems, one of which concerns the number of vacancies which exist in equilibrium in a solid.

A vacancy occurs when an atom in the solid leaves its normal lattice position. The atom must go somewhere else, as indicated in Fig. 10.1. Either the solid will expand a little, the extra atom being accommodated on a normal lattice site (case 1). Or it will sit, albeit rather uncomfortably, on an interstitial site (case 2). In the first case the vacancy will usually be formed initially at the surface and is then able to move through the solid by atomic diffusion. In the second case, both the vacancy and the interstitial defect, once formed, can diffuse through the solid. In fact the existence of vacancies is the mechanism by which atomic diffusion is able to take place, since only one atom has to move in order for a vacancy and a neighbouring atom to change places.

The question is: why do vacancies form, when clearly they require energy to do so? The answer concerns entropy, and the minimization of F. Certainly U must increase when a vacancy is formed. But at a high enough T, it is possible for this increase to be more than matched by a decrease in $(-TS)$, giving an overall lowering of $F(= U - TS)$. The vacancy (and its interstitial in case 2) are free to roam, and so give considerable disorder to the solid. Note that the whole discussion here could not be started without the free energy approach, since it is T and not U which is fixed.

To be specific let us develop a simplified model of case 1 vacancies (case 2 vacancies reappear as Exercise 10.5). Consider a simple solid of N atoms, which contains n vacancies at temperature T. The problem is to find how n varies with T. Suppose that

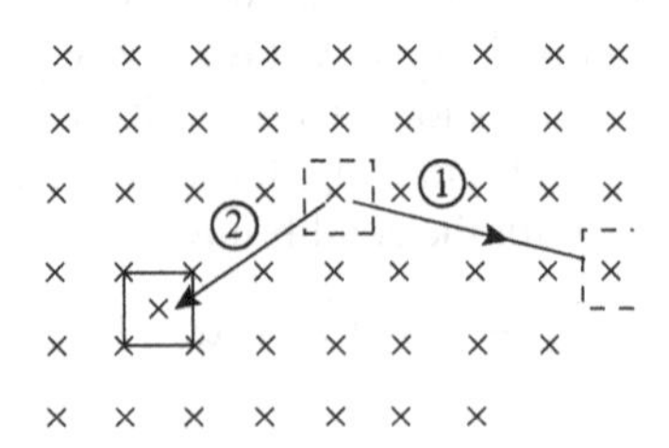

Fig. 10.1 Two types of vacancies which can occur in solids. Case 1: surface type. Case 2: interstitial type.

the formation energy of a vacancy is Φ, and that the vacancies are sufficiently dilute that they are weakly interacting. The increase in U due to the vacancies is then $n\Phi$.

The increase in entropy can be calculated in a manner similar to the isotopic disorder of section 10.1.1. The solid with n vacancies has $(N + n)$ lattice sites of which N are full and n are empty. Neglecting small surface effects, the number of possible arrangements is $(N + n)!/(N!n!)$. The increase in entropy due to the disorder of the vacancies is thus $k_B \ln[(N+n)!/N!n!]$. In practice the vibrational entropy of the solid will also increase slightly for secondary reasons, but we shall ignore this.

Following (10.7), the free energy increase when n vacancies are present is

$$\begin{aligned} F(n) &= U(n) - TS(n) \\ &= n\Phi - k_B T \ln[(N+n)!/N!n!] \\ &= n\Phi - k_B T[(N+n)\ln(N+n) - N\ln N - n\ln n] \end{aligned}$$

Equilibrium is given by the minimum F, and there are no constraints on n. Hence the required n (strictly n^*) satisfies $\mathrm{d}F/\mathrm{d}n = 0$, i.e.

$$\begin{aligned} 0 &= \mathrm{d}F(n)/\mathrm{d}n \\ &= \Phi - k_B T \ln[(N+n^*)/n^*] \end{aligned}$$

Bearing in mind that $n^* \ll N$ (since melting has not occurred), we may replace $(N + n^*)$ by N, to give the final result for the vacancy concentration

$$n^*/N \approx \exp(-\Phi/k_B T) \tag{10.11}$$

It is interesting that the answer is again just a simple Boltzmann factor. For most solids the value of formation energy Φ is about 1 eV, for example its value for copper is 1.4 eV. Equation (10.11) would give a vacancy concentration in copper of only about 1 in 10^{24} at room temperature (one vacancy per mole!), although in practice thermodynamic equilibrium would not be reached. Vacancy movement, i.e. diffusion, also requires an activation energy and disorder is frozen in at room temperature, another example of a metastable state. However, as the copper is heated the number of vacancies in equilibrium rises, and the time to reach equilibrium falls. By the melting point (1083°C), the number given by (10.11) is 1 in 10^5. For some substances, the number is even greater before melting occurs, and the increase in entropy of the solid can be observed directly as an additional heat capacity.

11

Phase transitions

Changes of phase are of great interest, not least because of their surprise value. In this chapter we examine how statistical physics can be used to help our understanding of some phase transitions.

11.1 TYPES OF PHASE TRANSITION

As mentioned earlier under our comments about the helium liquids (sections 8.3 and 9.2) phase transitions are commonly classified into 'orders'. In terms of the Gibbs free energy $G(G = U - TS + PV$ is the appropriate free energy since P and T are fixed in the phase change), a first-order transition occurs where the G surfaces for the two phases cross. The stable phase is the one with minimum G (compare the previous chapter's discussion of minimum F at fixed V and T), so that there is a jump in the first derivatives of G (i.e. S and V) at the transition. Hence 'first-order'. Furthermore supercooling and superheating effects can occur in a first-order transition, since the system can for a time be driven along the unstable (higher) branch of the G surface (Fig. 11.1).

However, this is *not* to be the topic of this chapter. Rather we shall be discussing second-order transitions. Here the changes in G at the transition are much more gentle. As the temperature is lowered, the system as it were eases itself gradually into a new phase which grows out of the first one. There is no superheating or supercooling; it is just that the G surface has a singularity at the transition. The singularity in a true second-order transition is such that S and V are continuous, but the second derivatives of G jump. Hence there is no latent heat (= change in S), but there is a jump in C (since $\mathrm{d}S/\mathrm{d}T$ changes). Exact second-order transitions are rare in practice, the best example being the transition to superconductivity of a metal in zero applied magnetic field. However, there are common examples of phase transitions which are close to second-order, namely the 'lambda transitions', like that in liquid ^{4}He (section 9.2). Other instances include transitions to ferromagnetism from paramagnetism, many transitions to ferroelectricity and some structural phase transitions from one solid phase to another.

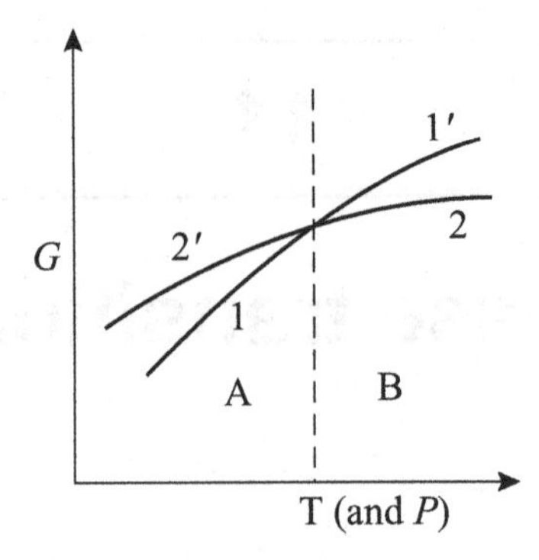

Fig. 11.1 The Gibbs free energy G around a first-order phase transition. In region A (of T and P) phase 1 is stable, whereas in region B phase 2 is stable. At the transition, jumps therefore occur in the first derivatives of G. Supercooling occurs when curve $2'$ is followed (an unstable situation) below the transition; superheating corresponds to curve $1'$.

A unifying concept for the understanding of all such transitions is that of an 'order parameter'. At high temperatures, the substance is disordered and the order parameter is zero. As one lowers the temperature through the transition, the order parameter suddenly starts to grow from zero, attaining its maximum value by the time the absolute zero is reached, a state of perfect order. For liquid ^{4}He and also for an ideal boson gas (of cold atoms, say), the order parameter is the superfluid density. For superconductivity it is the so-called energy gap. For ferromagnetism it is the spontaneous magnetization of the substance, i.e. the magnetization in zero applied field. And in the ordering of a binary alloy system (such as beta-brass, CuZn) with two types of lattice site, the order parameter characterizes which atoms are on which site.

In the sense that the onset of order in these systems is sudden, the transitions are often called 'order–disorder transitions'. The sudden onset of order implies a co-operative effect. In boson gases, the co-operation is forced by the (friendly) BE statistics. But in the other cases, the co-operation arises from interactions between the particles of the system. The importance of this chapter is that we can see how to apply statistical methods to at least one situation where the particles are *not* weakly interacting. In the next section, we shall take the transition from paramagnetism to ferromagnetism in a spin-$\frac{1}{2}$ solid as model example for these ideas.

11.2 FERROMAGNETISM OF A SPIN-$\frac{1}{2}$ SOLID

In section 3.1, we discussed the properties of a spin-$\frac{1}{2}$ solid, an ideal paramagnetic material, in which the spins are weakly interacting. In particular ((3.9) and Fig. 3.8), we worked out the magnetization M of the solid as the spins line up in a magnetic field B. The result was that if $B = 0$, then $M = 0$; but if $B \gg k_B T/\mu$ then M tends to the fully aligned value $M = N\mu$, the saturation magnetization. The spontaneous magnetization $M(0)$ of the solid (defined as the magnetization in zero applied magnetic field) is zero.

For the ideal paramagnetic solid, $M(0)$ will remain zero however far the temperature is lowered (see again section 3.1.3). However, what happens in any real substance is that a phase transition occurs at some temperature T_C, driven by interactions between the spins, to an ordered state. The simplest type of order, but certainly not the only type, is ferromagnetic ordering, in which the spins all tend to align in the same direction, giving a non-zero $M(0)$. (In all this chapter we ignore domain formation; $M(0)$ refers to the magnetization in a single domain, i.e. in a not too large sample or in a not too large region of a large sample.) The degree of order in the ferromagnetic state is characterized by an order parameter m defined by

$$m = M(0)/N\mu \tag{11.1}$$

Our main task is to understand how m varies with temperature.

The new ingredient of this chapter is to include in the argument a treatment of the interactions between spins. How is the energy of one spin influenced by the spin state of all of the others? That simple question has a complicated answer, unless approximations are made. One extreme approximation is an 'Ising model' in which the spins are assumed only to interact with their nearest neighbours. From the present viewpoint, that is still a complicated problem even in principle, although a start can be made at it using methods to be discussed in the next chapter. The other extreme approximation is to assume that all the other spins are of equal influence, however distant they are. This is called a *mean field approximation*, and we shall see that it gives a true second-order transition.

The mean field approximation is to assume that the one-particle states of each spin are, as usual, $+\mu B$ and $-\mu B$, but the value of B is changed. Instead of being equal to B_0, the applied magnetic field, the assumption is that we may write

$$B = B_0 + \lambda M \tag{11.2}$$

where λ is a constant characteristic of the substance. This follows since M is the magnetization of *all* the spins, so the term λM averages equally over all spins in the solid. The case $\lambda = 0$ recovers the ideal paramagnet of Chapter 3. The ferromagnetic situation corresponds to a large positive value of λ, a classic case of positive feedback as we shall now see.

11.2.1 The spontaneous magnetization (method 1)

We can directly give an expression for the magnetization of the solid, using the Boltzmann distribution with the energy levels implied by (11.2). Following (3.9) the answer is

$$\begin{aligned} M &= N\mu \tanh(\mu B/k_{\mathrm{B}}T) \\ &= N\mu \tanh[\mu(B_0 + \lambda M)/k_{\mathrm{B}}T] \end{aligned} \tag{11.3}$$

The spontaneous magnetization is obtained by setting the applied field $B_0 = 0$ in (11.3), giving a transcendental equation for $M(0)$. Replacing $M(0)$ by the dimensionless order parameter m, (11.1), we obtain

$$m = \tanh(mT_C/T) \tag{11.4}$$

where

$$T_C = N\mu^2\lambda/k_{\mathrm{B}} \tag{11.5}$$

Equation (11.4) can be readily solved graphically (or numerically) as in Fig. 11.2. Plotted as the x-variable on the graph is $x = mT_C/T$, so that (11.4) becomes $xT/T_C = \tanh x$. The solution is simply the intersection of the curve $m = \tanh x$ with the straight line $m = xT/T_C$. Since the slope of $\tanh x$ is unity close to $x = 0$, it is at once evident that:

1. When $T > T_C$, (11.4) has only one solution, namely $m = 0$. The spontaneous magnetization is zero, and the system is disordered.
2. $T = T_C$ is seen to be a definite transition temperature, in that below T_C other solutions are possible. T_C depends linearly on the mean field parameter λ. It is often called (for the ferromagnetic transition) the Curie temperature.
3. When $T < T_C$, the equation has three possible solutions. We shall see later that the $m = 0$ is now an unstable equilibrium. The solutions with $m \neq 0$ are the stable ones. Hence the solid develops below T_C a spontaneous magnetization with a definite magnitude, but with an indefinite direction.

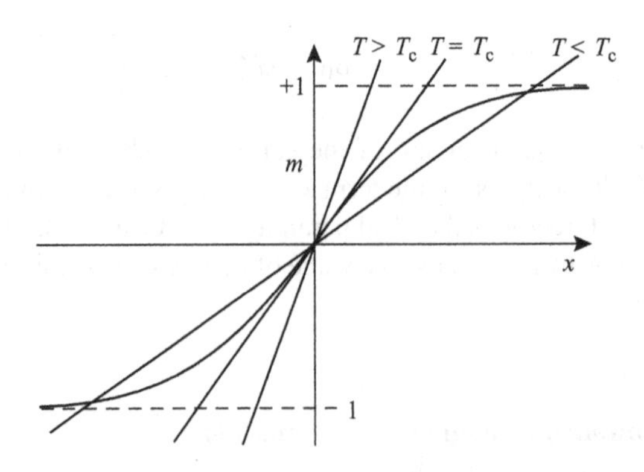

Fig. 11.2 Finding the spontaneous magnetization. The solution of (11.4) for m is given by the intersection of the tanh curve with the straight line. Straight lines are drawn for these different temperatures.

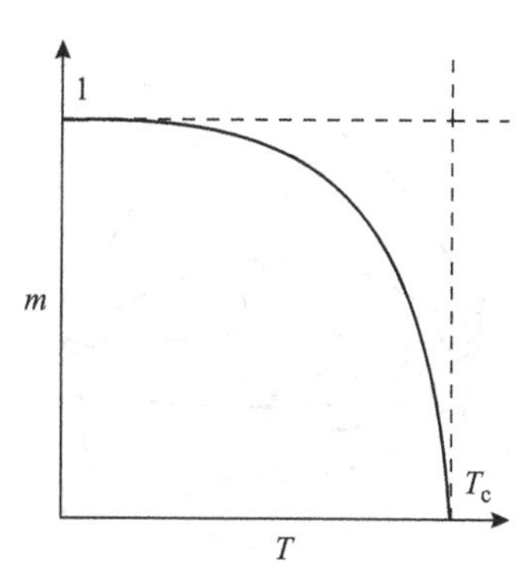

Fig. 11.3 Variation of the magnitude of the order parameter m with temperature T. Note that m varies from 1 at $T = 0$ to 0 at $T = T_C$. It remains zero above T_C.

The magnitude of m obtained by solving (11.4) is shown in Fig. 11.3, plotted as a function of T. The order parameter grows rapidly below T_C, reaching almost unity at $T_C/3$. One of the characteristic features of the mean field approximation is the total lack of order when $T > T_C$; the solid is effectively in a paramagnetic state in spite of the interactions between spins. However, before comparing with experiment, it is useful to rederive the result for $m(T)$ by another method.

11.2.2 The spontaneous magnetization (method 2)

When in doubt, minimize the free energy! That was the burden of Chapter 10, and we can do that here, as an alternative derivation of the variation of the order parameter m with temperature. To calculate F, we find both U and S as functions of m. The internal energy contribution due to the spins is simply their potential energy in the magnetic field. With the mean field assumption (11.2), this gives

$$\begin{aligned} U &= -\int_0 B \mathrm{d}M \\ &= -B_0 M - \lambda M^2/2 \end{aligned} \tag{11.6}$$

In zero applied field ($B_0 = 0$) the first term of (11.6) vanishes, and we have as a function of m

$$U = -(\lambda/2)(N\mu m)^2 \tag{11.7}$$

The entropy contribution from the spins can be worked out directly from (10.3). The fractions of the spins in the upper and lower states are given in terms of m by $P_2 = (1 - m)/2$ and $P_1 = (1 + m)/2$ respectively. [*Check:* $P_1 + P_2 = 1$, and

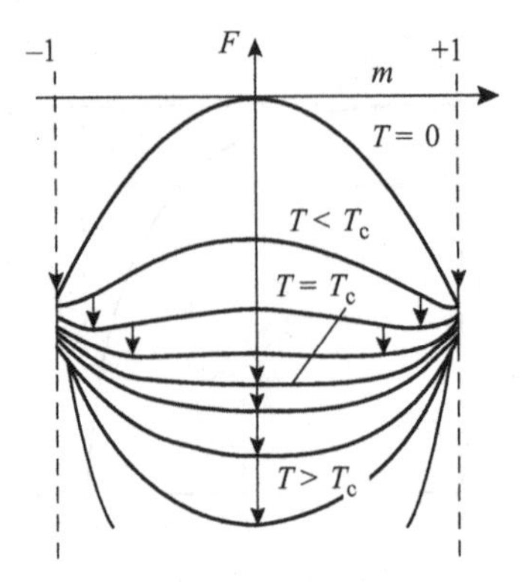

Fig. 11.4 Variation of F with m at various temperatures. The vertical arrows indicate the values of m for which F is a minimum at each temperature (compare Fig. 11.3).

$M = (P_1 - P_2)N\mu = N\mu m$, as required.] Hence

$$\begin{aligned} S &= -Nk_B(P_1 \ln P_1 + P_2 \ln P_2) \\ &= (Nk_B/2)[2\ln 2 - (1+m)\ln(1+m) - (1-m)\ln(1-m)] \end{aligned} \tag{11.8}$$

The free energy due to the spins $F = U - TS$ is now obtained by combining (11.8) and (11.7), together with the definition of T_C, (11.5), to give

$$\begin{aligned} F = (-Nk_B/2)\{T_C m^2 + T[2\ln 2 - (1+m)\ln(1+m) \\ - (1-m)\ln(1-m)]\} \end{aligned} \tag{11.9}$$

The stationary values of F are found by setting $\mathrm{d}F/\mathrm{d}m = 0$. It is a matter of elementary mathematics (try it!) to show that this leads to precisely (11.4), as it should. But we have learned much more from this exercise, in that we now know the full form of $F(m)$, as plotted in Fig. 11.4. We confirm that at $T > T_C$ there is just the one minimum at $m = 0$. When $T < T_C$, there are the two minima rapidly moving towards $m = +1$ and -1; and the stationary point at $m = 0$ is a maximum, an unstable position. It is also interesting to note the very shallow minimum at $T = T_C$, which indicates a very insecure equilibrium value of m; this gives an explanation of the large fluctuations often observed in the properties of a substance close to a transition point.

11.2.3 The thermal properties

The ordering of the spins is not just evident in the magnetization, but it will also give contributions to the thermal properties, for example to the entropy and to the heat capacity.

Since the variation of m with T is known (Fig. 11.3), the variation of S with T may be worked out from (11.8) and thence the heat capacity C from $C = T\,\mathrm{d}S/\mathrm{d}T$. The

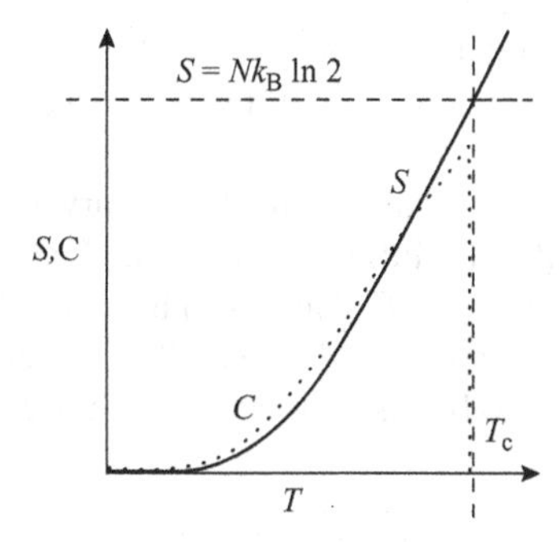

Fig. 11.5 Entropy and heat capacity in the mean field approximation. Full curve, entropy. Dotted curve, heat capacity (vertical scale × 0.6).

T-dependences of S and C in our mean field approximation are shown in Fig. 11.5. The entropy curve shows the sudden onset of order as T is reduced through T_C. Above $T_C, S = Nk_B \ln 2$ and the spins are completely disordered. Correspondingly, C vanishes above T_C, but shows a considerable anomaly over the temperature region below T_C in which S is changing.

11.2.4 The paramagnetic region

Notwithstanding the zero heat capacity in the paramagnetic ($T > T_C$) regime, we have already remarked upon the very shallow $F(m)$ curves close to the transition. Therefore, it is not surprising that the magnetic properties are affected above T_C. Although the *spontaneous* magnetization is zero, the response to an applied field is increased by the ferromagnetic interactions.

The magnetization in this region can be worked out readily as in Chapter 3. Equation (11.3) can be simplified (since B_0 and M are both small) by replacing $\tanh x$ by x to give

$$M = N\mu\,[\mu(B_0 + \lambda M)/k_B T]$$

and using the definition (11.5) of T_C this rearranges to

$$\frac{M}{B_0} = \frac{N\mu^2}{k_B(T - T_C)} \tag{11.10}$$

This should be compared with (3.10) for the ideal paramagnet. The new feature is the $-T_C$ in the denominator, which ensures that the 'paramagnetic' susceptibility does not follow Curie's law, but the so-called Curie–Weiss law. A graph of inverse susceptibility against T remains linear, but does not pass through the origin. Since $T_C > 0$ the susceptibility is everywhere increased by the interactions, and in fact diverges as the transition is approached – even a small applied field is enough to line up all the spins. This relates to the flat $F(m)$ curves, since the addition of an applied field B_0 to Fig. 11.4 (as in (11.6)) tilts the curves somewhat, and can move the minimum position well away from $m = 0$.

Incidentally, the same idea of tilting the $F(m)$ curves can be used in the *ferromagnetic* region to describe the influence of an applied magnetic field below T_C. One minimum becomes of lower F than the other, and the whole picture gives a good account of hysteresis curves (of M versus B_0) in a ferromagnetic material. The interested reader is referred to books on solid-state physics.

11.3 REAL FERROMAGNETIC MATERIALS

First a word of warning! We have outlined a simple theory for ferromagnetic interactions in a spin-$\frac{1}{2}$ solid.

Note: (i) In many cases the interactions in magnetic materials are antiferromagnetic, in other words neighbouring spins tend to be oppositely aligned (or to be at a non-zero angle). Ferromagnetism is not the only, and indeed not the commonest, situation. Nevertheless iron and nickel are not unknown materials, either. (ii) As in Chapter 3, extension to spins other than $\frac{1}{2}$ can readily be made, with no qualitative changes.

The mean field theory has notable successes in describing the transition from ferro- to para-magnetism. The nature of the spontaneous magnetization, the existence of a critical temperature T_C, the heat capacity anomaly and the magnetic properties are all described with reasonable accuracy. A true second-order transition (S continuous, but $\mathrm{d}S/\mathrm{d}T$ discontinuous) is predicted by the theory. How does all this compare with experiment in detail?

The first comment must concern the magnitude of the parameter λ. In (11.2) it seems that we have assumed that λM is a real magnetic field. And in some cases it might well be so; there will always be a true magnetic coupling between the dipole moments of adjacent spins. But in the materials we think of as ferromagnets (like iron for example), the physical origin of λ comes from far stronger effects, the magnitude of λM being several hundred tesla to give T_C above room temperature. This strong coupling arises from quantum influences on the electrostatic Coulomb forces between overlapping electrons on neighbouring atoms. Since electrons are identical spin-$\frac{1}{2}$ fermions, the overlap is strongly dependent on the relative spin of the two electrons, and hence the energy of one electron is influenced by the spin of its neighbours. This gives energy splittings as suggested by (11.2), but the origin is not a weak magnetic one, but is due to these much larger 'exchange energies'.

Next let us consider the detailed shape of the $m(T)$ variation (Fig. 11.3). Although agreement with experiment is fair, there are two points of detail which are incorrect. At low temperatures, the mean field approach somewhat overestimates m. This is because it ignores the existence of 'spin waves', essentially long wavelength periodic variations in magnetization. These are not included in our simple theory, which is based on long-range order only, i.e. on the spins of all ranges having an equal influence.

In practice spin waves do exist, and they are seen in the low-temperature region both in m and also as a small additional heat capacity.

The other region of interest is close to T_C. Here the mean field theory gives too small a value of m. The transition is somewhat more catastrophic than we have predicted. The theory very close to T_C suggests $m \propto (T_C - T)^{1/2}$, whereas reality is closer to $m \propto (T_C - T)^{1/3}$. The study of such 'critical exponents' is a very searching way of understanding phase transitions.

Correspondingly, the thermal properties show deviations from Fig. 11.5, and we do not have a true second-order transition. The heat capacity shows a lambda singularity, differing from the theory in two respects. Firstly (as with m) it is more sharp just below the transition, giving a much higher value as T_C is approached. And secondly there is an additional heat capacity just above the transition (the other half of the λ-shape). These effects are thought to arise from the existence of short-range order. Above T_C, although $m = 0$ identically, i.e. there is no long-range order, we still expect neighbouring spins to interact and perhaps form transient islands of local order. (We have already noted the very large fluctuations possible here.) The paucity of the mean field approach is that, as we have seen, it treats all spins equally, whereas in reality the interaction with closer neighbours will be stronger than that from distant ones. So treatments based on, say, an Ising model will describe these details more accurately. But that would also take us far beyond the scope of this chapter!

Finally (compare section 11.2.5) we can note that there is further evidence of the influence of short-range order in the magnetic properties above T_C. Until one is close to the transition, the form of (11.10), the Curie–Weiss law, is found to agree well with experiment. However, the parameter 'T_C' which enters it is found to be not the same as the actual ferromagnetic transition temperature, but rather a temperature a little higher (e.g. about 375°C in nickel, compared with a T_C of 358°C). Close to the transition it seems that M/B_0 is better described as $(T - T_C)^{-4/3}$, another example of mean field theory not getting the critical exponent quite right.

11.4 ORDER–DISORDER TRANSFORMATIONS IN ALLOYS

Before leaving the topic, we may note that the same type of mean field theory can be constructed for many other transitions. Here we just consider one other example, that of the phase transition which occurs in an alloy such as beta-brass CuZn.

This alloy contains equal numbers of Cu and Zn atoms. At high temperatures, the atoms lie on a body-centred cubic lattice in an entirely random arrangement. But below a transition temperature (about 460°C), the atoms start to form an ordered arrangement, with one type of atom at the corners of each unit cell and the other type at the centre of the cell. (A body-centred cubic structure can be thought of as consisting of two interpenetrating simple cubic sub-lattices, analogous to a two-dimensional centred square lattice, illustrated in Fig. 11.6.)

The transformation can be described by the mean field approximation, with much the same success and failure as in the ferromagnetic situation. The long-range order

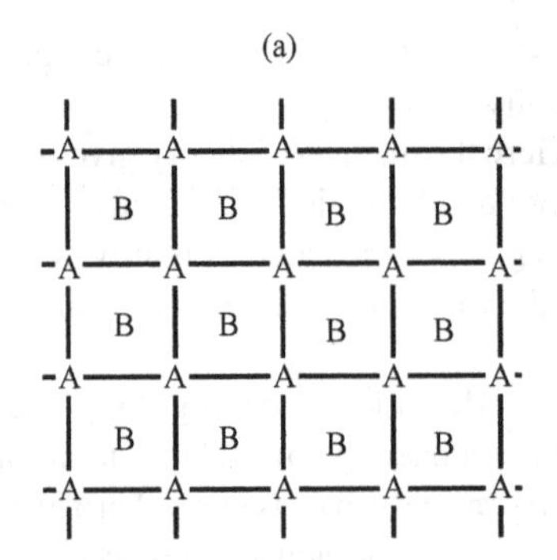

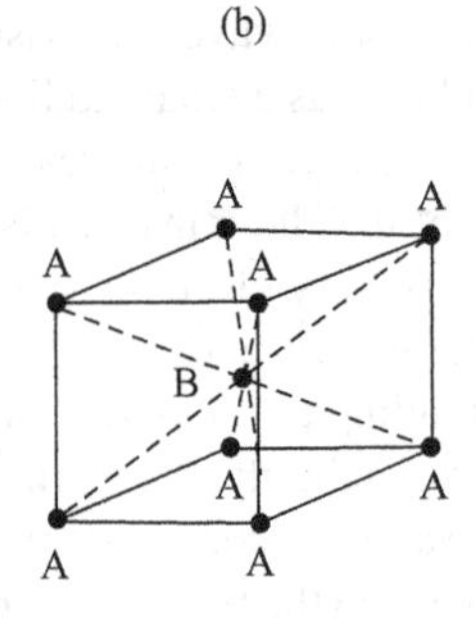

Fig. 11.6 Lattice sites in a crystal. The sketches show two (equivalent) types of lattice site, A and B, (a) in a square lattice in two-dimensional, and (b) in a three-dimensional body centred cubic lattice, as occurs in beta-brass.

parameter m can in this situation be defined as: $m = P_A - P_B$. Here P_A is the proportion of Cu atoms on A sites in the lattice (which is also the proportion of Zn atoms on B sites), and P_B is the proportion of Cu atoms on B sites. At high temperature, we expect $P_A = P_B = \frac{1}{2}$, so that $m = 0$. But in the fully ordered state we would have $m = +1$ or -1, depending on whether Cu or Zn occupies the A sites (another case of a 'spontaneously broken symmetry'!).

Whether the transition proceeds depends on the relative bond energy between like and unlike nearest neighbours. If it is favourable for neighbours to be unlike, we may expect a transition to occur. Writing V_{XY} as a bond energy between neighbouring X and Y atoms, we require $V = V_{CuCu} + V_{ZnZn} - 2V_{CuZn} > 0$. This parameter V plays the role of λ in the ferromagnetic situation. The mean field approximation in the present case is to assume that the number of bonds of each type is determined purely by the parameter m, assuming that the positions on the sub-lattices have no short-range correlations. It then follows (see Exercise 11.3) that the lattice internal energy can be written as $U =$ constant $-2NVm^2$ (compare (11.7)). The whole problem can then be solved as in section 11.2.

The graphs of section 11.2 may then be applied to this situation also. The ordering (i.e. $m(T)$) is easily followed by the appearance below T_C of extra lines in an X-ray powder photograph. The effects on the thermal properties are also readily measured, and there is an anomaly in C near the transition, associated with the entropy changes. As before, the anomaly in practice indicates a lambda transition, rather than the simple second-order transition of Fig. 11.5. Again, since the physical interaction is based on nearest neighbour interactions, it is not surprising that the mean field approach, with its neglect of short-range order, does not contain the full truth.

12

Two new ideas

In this chapter we shall explore briefly two concepts which indicate that statistical physics is not such a restrictive subject as might at first appear. The second gives clear pointers as to how we can start to deal with interacting particles. But before that, we shall consider connections between thermal equilibrium properties and transitions between states.

12.1 STATICS OR DYNAMICS?

Thermal equilibrium really is all about transitions, in spite of its static (and boring?) sound. As stressed in Chapter 1, our assembly samples effectively all its possible states over the time of an experiment, otherwise the whole concept of thermal equilibrium becomes misty. The idea of a dynamic equilibrium gives all sorts of connections between equilibrium and transport properties. In this section we explore just one such connection.

12.1.1 Matter and radiation in equilibrium

This is hardly a new topic, since it was introduced by Einstein. However, it is particularly important because it brings together the properties of matter and radiation in thermal equilibrium, along with an understanding of the emission and absorption of radiation by matter.

Consider a box which contains both radiation and a number of atoms, all in thermal equilibrium at temperature T. To be specific suppose the atoms are localized particles, which have two energy states, separated by Δ (they may have other energy states also, but these are irrelevant). Let there be N_2 atoms in the upper state, and N_1 in the lower state. One thing we have proved (Chapter 2) is that in equilibrium the ratio of the numbers of atoms in the two states is given by the Boltzmann factor

$$N_2/N_1 = \exp(-\Delta/k_B T) \qquad (2.26) \text{ and } (12.1)$$

What about the radiation? Thermal equilibrium is described in section 9.3. The photons occupy states of energy ε with an occupation function which is the modified

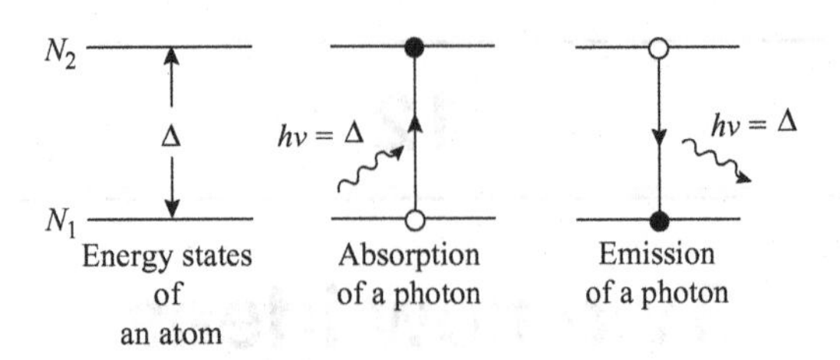

Fig. 12.1 Matter and radiation.

Bose–Einstein distribution

$$f(\varepsilon) = 1/[\exp(\varepsilon/k_B T) - 1] \quad \text{(9.9) and (12.2)}$$

Let us consider the mutual equilibrium between the atoms and the photons. Clearly there can be a strong interaction between the atoms and those photons which have an energy $\varepsilon(= h\nu) = \Delta$. There can be an absorption process (atom in state 1 + photon of energy $\Delta \rightarrow$ atom in state 2) which conserves energy; and an emission process which is the reverse (atom in state 2 $\rightarrow$ atom in state 1 + photon) (see Fig. 12.1).

Einstein's argument was to invoke the *principle of detailed balance*. This states that in equilibrium the average transition rates of absorption and emission processes must be equal (and this must be true for any transition and its inverse – hence 'detailed' balance). Now the upward transition rate, the number of absorption processes per unit time, is plausibly given by: $R(\text{up}) = N_1 f(\Delta) g(\Delta) X$. The process needs an atom in state 1, and it needs a photon of the right energy ($g(\varepsilon)$ is the photon density of states). The factor X describes the basic coupling strength, given the correct input of atom and photon.

What can we say about R(down), the emission rate? It requires an atom in the upper state, so it must be proportional to N_2. And that is enough to determine it fully! Detailed balance tells us that $R(\text{down}) = R(\text{up})$, and the equilibrium distributions ((12.1) and (12.2)) imply that

$$N_2/N_1\ [= \exp(-\Delta/k_B T)] = f(\Delta)/[1 + f(\Delta)]$$

Hence

$$R(\text{down}) = N_2[1 + f(\Delta)]g(\Delta)X \quad (12.3)$$

Equation (12.3) has a lot to reveal about the emission processes. The first is that the same coupling parameter X should be used for absorption and emission, an idea that should come as no surprise to anyone who has studied much quantum mechanics (the same matrix element is involved), but Einstein was inventing the subject! But of continuing interest is the $(1 + f(\Delta))$ factor in (12.3). The '1' term relates to what is called *spontaneous emission*, emission which will occur even if there are no other photons present. The treatment shows that it has a rate which is related both to the absorption rate and to the density of photon states into which the emission occurs. The $f(\Delta)$ term is responsible for *stimulated emission*. The emission rate is greatly increased into a photon state of the right energy, if there are already photons present in the state (those friendly bosons again!). That is the essence of laser action, in which atoms deliberately prepared to be in the upper atomic state and then de-excite by emitting photons into a cavity which supports a particular photon state. Hence the intense and coherent radiation.

Looked at from the viewpoint of statistical physics, the existence of stimulated emission is a necessity. In a strong radiation field, the stimulated term ensures that the dynamic equilibrium reached is one in which transitions up and down are frequent and $N_2 = N_1$, corresponding to infinite temperature. Disorder is rife, and the entropy is maximized. A laser needs $N_2 > N_1$, implying a negative 'temperature', i.e. a non-equilibrium prepared situation, one of lower entropy. Without the $f(\Delta)$ term in (12.3) one could cheat the second law; simply applying radiation would pump all the atoms up into the top state, achieving this desirable situation by equilibrium means.

12.1.2 Transitions with electrons

The principle of detailed balance may be used to establish the same sort of connection between Boltzmann statistics and Fermi–Dirac statistics. Consider as an example a metal which contains some magnetic impurities. The conduction electrons in the metal can be modelled as an ideal FD gas, which we know in thermal equilibrium will have the distribution of (8.2). The magnetic impurities can again be considered as a number of atoms with two energy states separated by Δ. As before, their equilibrium distribution is to be given by (12.1).

The transitions in question are inelastic scattering processes of an electron from the impurity atom. In an 'up' transition (of the atom, as previously) an atom in the lower state collides with an electron of energy $(\varepsilon + \Delta)$ leaving the atom in the upper state and the electron with energy ε. The reverse 'down' transition sees the atom go from state 2 to state 1, whilst the electron energy increases from ε to $(\varepsilon + \Delta)$.

Detailed balance should apply to every transition, whatever electron energy ε is involved. Since the electrons are fermions obeying the exclusion principle, the requirements for the up transition are, (i) an atom in state 1, (ii) a filled electron state of energy $(\varepsilon + \Delta)$, and (iii) an empty electron state of energy ε. Hence, in the spirit of the previous section, we would expect the transition rates to be

$$R(\text{up}) = N_1 f(\varepsilon + \Delta)[1 - f(\varepsilon)]g(\varepsilon + \Delta)g(\varepsilon)X$$

and (12.4)

$$R(\text{down}) = N_2 f(\varepsilon)[1 - f(\varepsilon + \Delta)]g(\varepsilon + \Delta)g(\varepsilon)X$$

where $f(\varepsilon)$ is the distribution and $g(\varepsilon)$ the density of states for electrons of energy ε. The factor X is again the appropriate coupling strength for the transition under examination. Now using detailed balance $R(\text{up}) = R(\text{down})$, and assuming

$N_2/N_1 = \exp(-\Delta/k_B T)$, it follows from (12.4) that the distribution function for the electrons must be of the form

$$[1 - f(\varepsilon)]/f(\varepsilon) = B\exp(\varepsilon/k_B T)$$

with B a constant (independent of ε). This rearranges to give

$$f(\varepsilon) = 1/[B\exp(\varepsilon/k_B T) + 1]$$

which is precisely the FD distribution (8.2) as indeed it should be. The argument of this section can be seen either as an alternative derivation of the FD distribution, or as a confirmation of the principle of detailed balance. We may further note that similar arguments can be made (with equally satisfactory results) for transitions of the atoms which involve, instead of electrons, either massive bosons or other Boltzmann particles.

12.2 ENSEMBLES – A LARGER VIEW

Statistical physics has many possible entry points. In this book, we have concentrated on one of the least abstract routes, that which concentrates on a piece of matter modelled by the (N, U, V) macro-state. The assembly under consideration consists of N weakly interacting particles, and the total energy U is fixed. Both of these limitations are essential for the method of Chapter 1 onwards to make sense.

But here is the new idea. Let us raise the scale of our thoughts, and apply the identical methods to a larger universe. Previously, we had an assembly of N identical particles with a fixed total energy U. Now we shall consider as 'ensemble' of N_A identical assemblies with a fixed total energy $N_A U$.

Why? Who wants to consider 10^{23} or so blocks of copper, when we are really interested in only one of them? The answer, of course, concerns averaging. We are interested in the average properties of one assembly, and this ensemble is one way of letting nature do the averaging for us! We can think of the ensemble as consisting of one assembly (whose thermodynamic properties we wish to know) in a heat reservoir provided by all the others, so that this is a way of modelling an assembly at fixed *temperature*, rather than the old method of fixed energy.

Furthermore, our statistical method applies immediately and precisely to this ensemble. The assemblies are separate macroscopic entities, so they are distinguishable 'particles' without a doubt. They can be placed in as weak contact as necessary, so the 'weakly interacting' limitation is unimportant. And we can avoid any doubts about numerical approximations by making N_A effectively infinite – after all, the ensemble is a figment of the imagination, not a blueprint for a machinist. Hence, the Boltzmann distribution (Chapter 2) applies.

For the assembly of particles, we derived the Boltzmann distribution

$$n_j = (N/Z)\exp(-\varepsilon_j/k_B T) \qquad (2.23)$$

with the partition function being

$$Z = \sum_j \exp(\varepsilon_j / k_B T) \tag{2.24}$$

the label j going over all one-particle states.

The parallel result for the ensemble involves the energy states of one assembly. Since an assembly is itself a large and complex entity, it has very many states of a particular energy. In fact, these are the *microstates* of Chapter 1; an assembly of energy E_j has $\Omega(E_j, V, N)$ such states. Since the allowed values of E are very close indeed, we usually adopt the density of states approximation and define $G(E)\delta E$ as the number of one-assembly states (= microstates) with energy between E and $E+\delta E$. The assembly partition function Z_A is defined as the sum over all one-assembly states of the Boltzmann factors, i.e.

$$Z_A = \sum_j \Omega(E_j, V, N) \exp(-E_j / k_B T) \tag{12.5a}$$

or

$$Z_A = \int_0^\infty G(E) \exp(-E/k_B T)\, dE \tag{12.5b}$$

The distribution is specified by the number $N(E)\delta E$ of the N_A assemblies in the usual energy range, or rather more usefully by the proportion $P(E)\delta E = N(E)\delta E / N_A$. The Boltzmann result then becomes

$$P(E) = N(E)/N_A = G(E) \exp(-E/k_B T)/Z_A \tag{12.6}$$

As before we may use (12.5) and (12.6) to calculate the thermodynamic functions for the ensemble – and hence for the average assembly. We write the total free energy as $N_A F$ and the total entropy as $N_A S$ (compare the definition U above). Thus the symbols U, S and F refer to the required average values per assembly. Following the 'royal route' of Chapter 2, we obtain for the average assembly

$$F = -k_B T \ln Z_A \qquad \text{(12.7) compare (2.28)}$$

Hence all the thermal properties of the assembly may be calculated from the microstates.

Several comments follow concerning the new viewpoint as expressed in (12.5), (12.6) and (12.7).

1. *Temperature.* The value of T which appears in the equations is determined from the properties of the whole ensemble (just as β in Chapter 2 was a property of the whole assembly). Therefore, as remarked above, when we concentrate on the behaviour of just one assembly we are discussing its properties when in contact with a large (conceptually infinite) heat reservoir at this temperature T.

2. *Internal energy.* Although T is fixed by the ensemble, the internal energy of any one assembly is not. The energy of an assembly can have any value E, with a probability given by (12.6). The one thing we can be sure about is that the *average* value of E is U, for that was one of the starting points. However, (12.6) contains the full statistical information about E, and that we shall now examine in greater detail.
3. *The distribution function $P(E)$.* For assemblies of any reasonable size, it turns out that $P(E)$ is a very strongly peaked function indeed around its maximum value (which thus effectively equals the mean value U). This comes about because the assembly 'density of microstates' function $G(E)$ is a rapidly rising function of E, whereas the exponential factor in (12.6) is a rapidly falling function. Consider (without working the problem out in detail) a specific case, where the answer is already well known from Chapter 6. This is where each assembly consists of N dilute gas particles, to which MB statistics apply. Suppose that $G(E) \propto E^n$. Then (equation (12.6)) $P(E) \propto E^n \exp(-E/k_BT)$. The maximum is given by $dP/dE = 0$, i.e. (check it!) when $E = nk_BT$. In our example we know that $U = 3/2Nk_BT$, so the index n must be equal to $3N/2$. This is a fantastically high power for the rising factor. Correspondingly the falling exponential factor has an exponent $-3N/2$, quite enough to frighten any normal calculator.
4. *Fluctuations.* Continuing the previous point, $P(E)$ also enables us to work out the probable fluctuations of E around the mean value U. The sharpness of the maximum is readily characterized from the second derivative d^2P/dE^2 at the maximum. A Taylor expansion of $P(E)$ about the maximum shows that a useful measure of the width ΔE of the curve is given by $d^2P/dE^2 = P(U)/(\Delta E)^2$. For the above example, this gives $\Delta E = n^{1/2}k_BT$, or more usefully the fractional width $\Delta E/U = 1/\sqrt{n} \approx 1/\sqrt{N}$. The relative fluctuations in E are large for a small assembly at a fixed temperature, but reduce as about $1/\sqrt{N}$ for a large assembly at a given T.
5. *The same answers for the same problems.* Naturally since no new assumptions have been made, the problems we have already tackled will still have the same solutions attacked by the new method. For instance let us consider an assembly of N distinguishable particles as in Chapter 2. We can calculate Z_A by one of two methods.
   (i) We can use the ideas of the factorization of a partition function elaborated in our discussion of diatomic gases (section 7.1). Since the particles are weakly interacting, the microstate simply specifies the state of each individual particle in turn, and the various one-particle energies (all N of them) simply add to give E, the assembly energy. Therefore the assembly partition function (12.5a) is given simply by

$$Z_A = Z^N$$

each particle contributing a factor Z (equation (2.24)). We can immediately check that this is correct, since $F = -k_BT \ln Z_A$ (equation (12.7)) $= -Nk_BT \ln Z$ (identical to (2.28)).

(ii) We may also calculate Z_A using (12.5a) together with the multinomial theorem (Appendix A). Because the coefficients $t(\{n_j\})$ are identical (for an assembly of distinguishable particles) to the multinomial coefficients, it is not too hard to prove that the sum in (12.5a) can be performed explicitly; and that the result again is Z^N.

Method (ii) is worth examining in addition to method (i), since when we come to gases method (i) is not available. The gas particles are competing for the same states, so Z_A cannot factorize into a factor from each particle. However, method (ii) gives us the answer for a dilute (MB) gas. (The calculation of Z_A for FD and BE gases requires yet further work, which we shall not pursue here.) The values of $t(\{n_j\})$ for the MB gas simply differ by that factor $N!$ from those for the distinguishable Boltzmann particles. Therefore the summation goes through unscathed, except for this constant factor throughout, to give

$$Z_A = Z^N/N!$$

for the assembly partition function. This expression, when put into equation (12.7), gives $F = -Nk_BT \ln Z + k_BT \ln N!$, identical to the result of Chapter 6 (6.16).

6. *A way into new problems.* The importance of the new method is that a wider range of assemblies can now be studied, at any rate in principle. The three equations (12.5), (12.6) and (12.7) outline a programme which should work for an ensemble of any assemblies. The equations make no restrictions on 'what is inside each box', the only requirement being that we have in mind a large number of identical boxes. So the assembly could contain interacting particles, and still the method would be the same. Naturally it becomes mathematically difficult, but at least the physics is clear. The computational problem is immense, as soon as the energy E depends on interactions between the 10^{23} particles. Nobody can yet fully work out the properties of a liquid from first principles. However, some progress can be made. For example, the small deviations from the ideal gas laws for real gases (which are only slightly non-ideal) can be related to the interaction potential between two gas particles, using expansion methods effectively in powers of the interaction. We shall use these ideas further in Chapter 14. Another tractable problem is that of the Ising model, mentioned as a (more physical?) alternative to the mean field approximation of Chapter 11. Here we have a solid in which only nearest neighbours interact, so that the expression for E is not too desperate. In fact it is exactly soluble in two dimensions, and satisfactory numerical results can be obtained in three dimensions.
7. *Nomenclature.* You know the methods, so why not learn their names? (At least it helps to impress your friends!) The ensemble of this chapter, with a probability function for E given by (12.6), is called the *canonical ensemble*. The earlier method of the book can also be thought of as involving an ensemble (see section 1.3); but now each assembly of the ensemble had the same energy U. In the language of (12.6), $P(E)$ for the ensemble is a delta-function at energy U. This is called the *micro-canonical ensemble*.

8. *New horizons.* Finally, we can point out that the degree of abstraction need not end here. In the canonical ensemble the restriction that U is fixed is removed, in comparison to the microcanonical ensemble. But the assemblies are still constrained to have a given number N of particles. For some purposes this is quite unphysical. For example consider the properties of a litre of air in the middle of the room. The properties of this 'open assembly' are thermodynamically well defined and calculable. We can describe its properties in statistical physics with a *grand canonical ensemble* which consists of an array of these open assemblies. Now there is not only energy exchange between assemblies, but there is also particle exchange. What happens now is that the whole ensemble defines a chemical potential μ, determined by the total number of particles in the ensemble, just as T in the canonical ensemble was determined by the total energy ($N_A U$) of the ensemble. In the grand canonical ensemble, an assembly may have any number of particles, but, as one might anticipate, there is again a very sharp probability function $P(E, N)$ which ensures that E and N are in practice well defined. As with the canonical ensemble, the grand canonical ensemble solves all our old problems equally well. But in addition it opens new techniques, essential, for example, for discussing chemical reactions and equilibrium and for a study of solutions. We develop these ideas further in the following chapter.

13

Chemical thermodynamics

In this chapter we will build on ideas introduced earlier in the book, notably the idea of chemical potential, and its relevance to a discussion of phase transitions (following section 11.1). We shall extend the concept of the ensembles (section 12.2) to discuss the grand canonical ensemble and an approach to open systems. These ideas come together to discuss simple chemical reactions and conditions for chemical equilibrium.

13.1 CHEMICAL POTENTIAL REVISITED

We met the chemical potential μ at the end of Chapter 2. The idea of chemical potential also has appeared frequently, notably in the discussions of FD gases (Chapter 8, where it was called the Fermi energy) and of real BE gases (Chapter 9, where it was hidden in the B parameter). (The name chemical potential was not always used, perhaps because of the physicist's reluctance to mention chemistry?)

In statistical physics, the flavour of chemical potential is that it is a *potential for particle number*. It is intimately associated with α, the Lagrange multiplier introduced in conjunction with the number condition $\sum n_i = N$. In fact as we have seen it is related by

$$\alpha = \mu / k_B T \tag{13.1}$$

In thermal physics (see, for example, *Thermal Physics* by Finn, chapter 10), chemical potential is defined as being an appropriate number derivative of an energy function. Specifically, in terms of internal energy U it is defined as

$$\mu = \left(\frac{\partial U}{\partial N} \right)_{S,V}$$

Putting together $dU = TdS - PdV + \mu dN$ with the definition of the Helmholtz free energy $F = U - TS$ we obtain

$$\mu = \left(\frac{\partial F}{\partial N} \right)_{T,V}$$

as used in Chapter 2. Stirring in the definition of the Gibbs free energy $G = F + PV = U - TS + PV$ we obtain $\mathrm{d}G = -S\mathrm{d}T + V\mathrm{d}P + \mu\mathrm{d}N$ and hence an equivalent definition of μ is:

$$\mu = \left(\frac{\partial G}{\partial N}\right)_{T,P}$$

This last result is particularly useful, since temperature and pressure are *intensive* parameters, whereas number alone in $G(T,P,N)$ is *extensive*. This means that if we double the number of particles present of a substance at constant T and P, in order for G to double, it is necessary simply that G and N are proportional. In other words it follows immediately that

$$\mu = G/N \tag{13.2}$$

and we see that another way of looking at the chemical potential is that it is the Gibbs free energy per particle.

This idea immediately relates to the discussion of phase transitions in Chapter 11. Consider again a pure (one-component) substance which can exist in two phases, labelled 1 and 2. We wish to decide which of the phases is the stable one in equilibrium at a particular pressure and temperature. The answer is easy. As always, the equilibrium condition is that which minimizes free energy. Suppose that we have a total N particles, of which N_1 particles are in phase 1 and N_2 in phase 2, so that $N = N_1 + N_2$. Using (13.2), we write for the total free energy of the system

$$G = G_1 + G_2 = N_1\mu_1 + N_2\mu_2$$

where μ_1 and μ_2 are the chemical potentials in the two phases. It is now clear that which phase is stable is determined by the magnitude of the two chemical potentials.

If $\mu_1 > \mu_2$, then G is lowest when $N_1 = 0$ and all the substance is in phase 2, the phase with the lower chemical potential. On the other hand, if $\mu_1 < \mu_2$ then $N_2 = 0$ and the stable phase is phase 1. In other words, particles move from high μ to low μ and it is seen that chemical potential is indeed a 'potential for particle number'.

If on the other hand $\mu_1 = \mu_2$, then G is the same whatever the particle number ratio N_1/N_2. This is the case of phase equilibrium. An indeterminate mixture of the two phases can be present. It is now worth turning back to Fig. 11.1, which shows a graph of G for the two phases. Essentially this is just the same as a graph of G/N, i.e. μ, for the two phases; and the above commentary describes precisely the equilibrium condition.

It is worth remarking here that the equality of chemical potential in an equilibrium system is a very powerful tool. As an example, it enables one to discuss a whole range of problems in semiconductor physics. In this case the appropriate potential is the so-called electrochemical potential, a sum of electric potential energy of an electron and the chemical potential. (Conduction electrons are charged and have their potential energies changed by an electric potential, whatever their kinetic energies). The direction of electron motion is determined by the gradient of this electrochemical potential. In equilibrium the electrochemical potential (often called the Fermi level in this context, as in Chapter 8) of the electrons becomes the same throughout the material. In an inhomogeneous material, this is achieved by small-charge transfers between the different regions, an idea that is fundamental to our understanding of the transistor and of a host of other devices.

13.2 THE GRAND CANONICAL ENSEMBLE

In order to discuss the properties of open systems, it is useful to look further at the concept of the grand canonical ensemble introduced in section 12.2. The idea is to develop a technique for a system in which particle number is not fixed from the outset, but is determined by the system being open to a 'particle bath' with a specific chemical potential μ.

This may be visualized as, say, being interested in a litre of air in the middle of the room. This system of interest, the particular litre in the room, has well-defined thermal properties, even though its energy U and its particle number N are not constants. Rather, U is controlled by the temperature T, a property of the whole ensemble. As discussed for the canonical ensemble in section 12.2, U has a probability distribution with a very sharp peak. We can say that *temperature is a potential for energy*. Similarly, N has a strongly peaked probability distribution determined by the chemical potential μ, again a property of the whole of the ensemble (room). As stated in section 13.1, μ *is a potential for particle number*.

This approach gives the right flavour, but is not enough to develop a quantitative description of an open system. A little mathematical imagination is required, and this we now discuss.

13.2.1 The new method

Method 1. The basic method adopted so far in this book centres around the following equation:

$$\frac{\partial \ln t}{\partial n_j} + \alpha + \beta\varepsilon_j = 0 \tag{13.3}$$

This appeared in section 2.1.5 (compare (2.11), following (2.7)), and in section 5.4. It is a statement of the Lagrange multiplier approach. Equation (13.3) is the recipe for finding the most probable distribution. It is designed to pick out the maximum value of t subject to the two restrictive conditions $\sum n_j = N$ and $\sum n_j\varepsilon_j = U$. It is a method for dealing with any type of assembly of weakly interacting particles, localized or gaseous, so long as the number of microstates t $(\{n_j\})$ for a particular distribution is appropriately expressed.

Method 2. But (13.3) can also be expressed in a way that relates to the canonical ensemble, introduced in Chapter 12. It is also a recipe designed to guarantee that

$t\exp(\beta U)$ is a maximum, subject now to one restrictive condition $\sum n_j = N$. Using the Lagrange method with the one condition (undetermined multiplier α) gives precisely (13.3). The link with the canonical ensemble approach is evident when we recognize that, within the usual approximations of large numbers, this view is adjacent to that based on the assembly partition function Z_A of (12.5). We simply need to approximate $\Omega(U, V, N)$ by its largest term t and to identify the physical meaning of the β multiplier, i.e. to set $\beta = -1/k_B T$. Equation (13.3) can then be seen to be effectively picking out the maximum term in Z_A. This is almost equivalent to evaluating Z_A itself, since there is a very sharp peak indeed as discussed earlier (note 3, following (12.5)).

Method 3. The interesting way of opening out (13.3) is now taken to one final stage. The equation certainly also gives the recipe to identify the distribution which makes the function $t\exp(\alpha N)\exp(\beta U)$ an unconditional maximum. And unconditional maxima are good news mathematically. Hence, the imaginative approach to which this argument leads is to:

1. Define a *grand partition function* Z_G by the expression

$$Z_G = \sum_{k,l} \Omega(E_k, V, N_l)\exp(\mu N_l/k_B T)\exp(-E_k/k_B T) \tag{13.4}$$

   where the sum goes over all energies (labelled E_k in this chapter) and all particle numbers (labelled N_l) from zero to infinity.
2. Note that the maximum term in (13.4) will be recovered from (13.4) if we again equate t and Ω, and give the physical identification to both multipliers α and β.
3. Suggest that a new method to describe equilibrium should be based on the full sum in (13.4), when this is convenient, rather than just on the largest term in the peaked distribution. We note that this method has all the attributes to describe open systems, since the construction of Z_G starts off by assuming that T and μ are known and that U and N are to be determined from them, by identifying the maximum term in Z_G or (as is equivalent in practice) by using the terms in Z_G as statistical weights in any thermodynamic averaging process.

13.2.2 The connection to thermodynamics

This is straightforward, but important.

Method 1 (Microcanonical ensemble). The connection is via (1.5), $S = k_B \ln \Omega$, appropriate to the set variables U, V and N. Hence

$$TS = k_B T \ln \Omega \simeq k_B T \ln t^*$$

This is the basis on which most of this book so far is based.

Method 2 (Canonical ensemble). The given variables are now T, V and N. As discussed in section 12.2, we define an assembly partition function Z_A (equation (12.5)) as

$$Z_A = \sum_k \Omega(E_k, V, N) \exp(-E_k/k_B T)$$

Using the maximum term approximation (i.e. removing the sum and setting $E_k = U$) then gives

$$k_B T \ \ln Z_A = k_B T \ \ln \Omega - U = -F$$

as already derived (equation (12.7)).

Method 3 (Grand canonical ensemble). The starting parameters are now T, V and μ. Following the definition of Z_G in (13.4), and again using the maximum term approximation, we now derive the following result:

$$k_B T \ \ln Z_G = k_B T \ \ln \Omega + \mu N - U = TS + G - U = PV \tag{13.5}$$

Here, we have used the identification of μN with G as discussed above in section 13.1. The result is that the appropriate thermodynamic energy function is simply PV.

Since it is somewhat unfamiliar as a thermodynamic function, it is worth stressing that PV as a function of T, V and μ is a thoroughly useable and useful idea. As discussed in section 13.1, since $G = F + PV = U - TS + PV$ and also, (13.2), $G = N\mu$, we may write $PV = N\mu - U + TS$. Since $\mathrm{d}U = T\mathrm{d}S - P\mathrm{d}V + \mu \mathrm{d}N$ we have immediately

$$\mathrm{d}(PV) = S\mathrm{d}T + P\mathrm{d}V + N\mathrm{d}\mu \tag{13.6}$$

Hence a knowledge of Z_G enables (using (13.5) and (13.6)) an effective and direct route to the determination of other thermodynamic quantities, such as S, P and in particular N, the (equilibrium) number of particles in our open system.

13.3 IDEAL GASES IN THE GRAND ENSEMBLE

It is instructive to rederive some of the results for ideal gases using this new method, to indicate the power and generality of this somewhat more sophisticated approach to statistical physics. The power comes about since, as we shall now see, the absence of restrictions on N and U makes it easy to evaluate Z_G for an ideal gas.

13.3.1 Determination of the grand partition function

We shall label the one-particle states by j, and their corresponding energies ε_j. For an ideal gas these energies depend on V only (fitting waves into boxes!). Note that in this treatment it is not necessary (or even convenient) to group the states together (the i notation of section 5.1). Rather we discuss the occupation of *individual* one-particle states (hence the j notation as Chapter 1). For compactness, we choose to define for each such state a quantity γ_j defined as

$$\gamma_j = \exp[(\mu - \varepsilon_j)/k_B T] \tag{13.7}$$

We may note that, with the expected physical identifications with our earlier approach, this is the same as $\gamma_j = \exp(\alpha + \beta\varepsilon_j)$, a frequently occurring quantity.

Consider one particular microstate of the assembly. Suppose that, in this microstate, state j is occupied by n_j (identical, gaseous) particles. The microstate thus has a total $\sum n_j = N_l$ particles and a total energy $\sum n_j\varepsilon_j = E_k$, where the sums go over all the one-particle states j. The contribution of this one microstate to Z_G (see (13.4)) is $\exp(\mu N_l/k_B T)\exp(-E_k/k_B T)$, i.e. it is simply equal to

$$\prod_j \gamma_j^{nj} \tag{13.8}$$

verified by substitution. There is one such term for every microstate of the grand assembly; and there are no restrictions whatever on the n_j since all N_l and E_k are to be included in the sum. To obtain Z_G we need to sum over all possible microstates, a task which sounds daunting but turns out to be amazingly easy.

The easy answer to the sum is different, depending on whether the identical gas particles are fermions or bosons, and we must treat the two cases separately.

Fermi–Dirac. Here, the Pauli exclusion principle (antisymmetric wavefunction) tells us that only two occupation numbers are possible. We must have $n_j = 0$ (empty states) or $n_j = 1$ (full states). Hence all the γ factors for each in (13.8) must either be $\gamma^0 = 1$ for empty states or $\gamma^1 = \gamma$ for full states. There are no total number restrictions, so that the sum over all microstates can therefore be written as

$$Z_G(\text{FD}) = \prod_j (1 + \gamma_j) \tag{13.9}$$

The appearance of 1 and γ in every bracket in this product ensures that every allowable microstate of the form prescribed by (13.8) is included in (13.9), and that it is included once only.

Bose–Einstein. The same technique also works for the BE case. The only difference is that now all occupation numbers are possible, i.e. n_j can take any integer value, and there are many more microstates which must be counted. The answer, however, is easy

$$Z_G(\text{BE}) = \prod_j (1 + \gamma_j + \gamma_j^2 + \gamma_j^3 \ldots) \tag{13.10a}$$

A particular microstate is defined by the number of particles n_j in a state j, for every one-particle state. And in (13.10a), this corresponds to picking out the term γ_j^{nj} from the jth bracket. This is not yet the simplest expression for Z_G, however, since the infinite sum in each bracket can readily be performed. We have already met sums of this sort in our discussion of an assembly of harmonic oscillators in section 3.2.1. The summation is $(1 + \gamma + \gamma^2 + \gamma^3 \ldots) = (1 - \gamma)^{-1}$. Hence we obtain the final result

$$Z_G(\text{BE}) = \prod_j (1 - \gamma_j)^{-1} \tag{13.10b}$$

Maxwell–Boltzmann limit. Having said that there are two cases, we now follow precedent to consider a third! The point is that we can obtain and use an even simpler expression for Z_G in the MB limit of either form of statistics, FD or BE. We already know that the two statistics tend to the same limit for a dilute gas, i.e. one in which the occupation of the states is very sparse. As usual, the FD and BE cases tend to the limit from opposite sides. In the dilute limit, one expects the exclusion principle to become an irrelevance, since the probability of double occupation is always negligible. The appropriate expression is obtained from a compromise between (13.9) and (13.10a) when all the γ_j are small, i.e. $\ll 1$. It is to replace either equation by

$$Z_G(\text{MB}) = \prod_j \exp(\gamma_j) \tag{13.11}$$

The compromise involved is seen when one recalls the power series, appropriate for small γ

$$\exp \gamma = 1 + \gamma + \gamma^2/2! + \gamma^3/3! \ldots$$

13.3.2 Derivation of the distributions

The distribution is defined as a filling factor for a one-particle state. It tells us the number of particles per state on average in thermal equilibrium. Earlier we defined the distribution f_i as the fractional occupation of states with energy ε_i. The definition is even more straightforward in the grand assembly, since we now consider the states individually. Thus the fractional occupation of state j is precisely the thermal average of n_j. We denote this average by the symbol $\overline{n_j}$, which has exactly the same interpretation as the earlier f_i. As postulated at the end of section 13.2.1, the thermal average is obtained using the relative weights of each term in the expression for Z_G.

Fermi–Dirac. The expression for Z_G is (13.9), where the product goes over all one-particle states j. Now let us work out the average occupation of one of those states, which at the risk of confusing the reader we shall label as i. (There are not enough letters in the alphabet; however this notation has the merit that the old f_i should look identical to the newly derived $\overline{n_i}$.) Each one of the many (infinite!) microstates which make up the grand canonical ensemble corresponds to a particular term in the extended product obtained by multiplying out (13.9). And the statistical weight of each microstate is simply the magnitude of this term, as postulated above. We need to examine the role of the $(1+\gamma)$ factor for which $j = i$ in this expansion. If our state i is full ($n_i = 1$), then we have taken the γ_i from the bracket; if the state is empty we have taken the 1. The thermal average required is thus equal to the sum of the terms in which $n_i = 1$ divided by the total sum Z_G of all the terms. Hence

$$\overline{n_i} = \frac{\gamma_i}{(1+\gamma_i)} \times \frac{\prod_{j\neq i}(1+\gamma_j)}{\prod_{j\neq i}(1+\gamma_j)}$$

The contribution for the 'other' states (those with $j \neq i$) conveniently factors out. We are left therefore with

$$\overline{n_i} = \frac{\gamma_i}{1+\gamma_i} = \frac{1}{\gamma_i^{-1}+1} = \frac{1}{\exp[(\varepsilon_i - \mu)/k_B T]+1}$$

as expected and hoped, in agreement with (5.13). This is the Fermi–Dirac distribution.

Bose–Einstein. Here the same sort of technique works, in that the same cancellation of the factors from the $j \neq i$ states takes place. However we must now allow for all occupation numbers in the state i. Looking at the bracket for $j = i$ in (13.10a), we recall that the term 1 corresponds to $n_i = 0$, the term γ_i to $n_i = 1$, the term γ_i^2 to $n_i = 2$ and so on. Therefore the expression for the distribution is

$$\overline{n_i} = \frac{\gamma_i + 2\gamma_i^2 + \cdots}{1+\gamma_i+\gamma_i^2+\cdots}$$

This expression looks somewhat intractable until we take a hint from the summation of the denominator, as in the transition from (13.10a) to (13.10b). We write $F = (1-\gamma)^{-1} = 1+\gamma+\gamma^2+\ldots$ as before. Differentiating both forms of F, we obtain

$$\frac{dF}{d\gamma} = (1-\gamma)^{-2} = 1+2\gamma+3\gamma^2+\ldots$$

Hence the expression for the distribution becomes

$$\overline{n_i} = \frac{\gamma_i dF/d\gamma_i}{F} = \frac{\gamma_i}{1-\gamma_i} = \frac{1}{\gamma_i^{-1}-1} = \frac{1}{\exp[(\varepsilon_i - \mu)/k_B T]-1}$$

This is the Bose–Einstein distribution of (5.13).

Maxwell–Boltzmann limit. Finally we may note without further ado that, since in the MB limit all $\gamma_i \ll 1$, either of the two distributions tend to the even simpler result: $\overline{n_i} = \gamma_i = \exp[(\mu - \varepsilon_i)/k_B T]$, the Maxwell–Boltzmann distribution as expected.

13.3.3 Thermodynamics of an ideal MB gas

The gas laws for the dilute (MB) ideal gas can now be rederived in short order from the grand canonical ensemble approach. In (13.11) we have found the appropriate expression for Z_G, the grand partition function. We can now substitute Z_G into the basic equation, (13.5), to find PV and hence other thermodynamic quantities. For FD and BE gases, the same method works well, but the mathematics get involved. However for an ideal gas in the MB limit, the answers are elementary and immediate (so we may as well do it!).

Equation (13.5) tells us that $PV = k_B T \ln Z_G$. The great simplification for the MB limit is that $\ln Z_G$ works out so easily. Taking the logarithm of (13.11) we obtain

$$\begin{aligned} \ln Z_G &= \sum_j \gamma_j = \exp(\mu/k_B T) \sum_j \exp(-\varepsilon_j/k_B T) \\ &= \exp(\mu/k_B T) Z \end{aligned} \tag{13.12}$$

The result is alarmingly simple. The *logarithm* of the grand partition function is a factor $\exp(\mu/k_B T)$ times $Z(V, T)$, the ordinary 'sum-over-states' partition function, first introduced in Chapter 2 (equation (2.24)) and which played a central role in our earlier discussion of the MB gas in Chapter 6 (e.g. section 6.1). Z is determined only by the temperature (energy scale $k_B T$) and by the quantum states of one particle, involving the box size V through the allowed quantum energies of the particle. We recall from Chapters 6 and 7 that for a spinless monatomic gas $Z = V(2\pi M k_B T/h^2)^{3/2}$. For a more complicated perfect gas, this expression, since it derives from translational kinetic energy only, must be multiplied by another factor Z_{int} to allow for the other internal degrees of freedom (spin, rotation, vibration etc). For a given gas, Z_{int} for rotation and vibration varies with temperature only (see Chapter 7), whilst for spin it will give a constant factor G. Thus for any perfect gas we have

$$Z(V, T) = \text{const } VT^n$$

where $2n$ is the number of degrees of freedom excited.

Let us now look at the thermodynamic functions. In the grand canonical ensemble approach, we are starting off by fixing μ, V and T. That enables us to evaluate Z_G and hence $\ln Z_G$ as in (13.12) above, and thus

$$PV = k_B T \exp(\mu/k_B T) Z(V, T) \tag{13.13}$$

Equation (13.6) now gives the prescription for calculation of other thermodynamic quantities. For instance S, P and N are obtainable from it by simple partial differentiation. We shall not labour over S here, but this approach would directly rederive the Sackur–Tetrode equation (6.15) for an ideal monatomic gas, and it would lead us into the heat capacity discussions of Chapter 7, with the index n appearing from differentiation with respect to T. Any further derivation of P is unnecessary here, since the job is done already in (13.13). But what about N, the average number of gas molecules in our open system? From (13.6) we see that the answer is

$$N = \left(\frac{\partial(PV)}{\partial \mu}\right)_{T,V}$$

Since μ appears only once in (13.13), the result is immediately accessible, namely

$$N = \exp(\mu/k_B T)Z \tag{13.14}$$

There are two important features of (13.14):

1. In combination with (13.13), it leads to the relation $PV = Nk_B T$ for the ideal MB gas. As expected, the ideal gas equation of state is valid for an open system (where N is determined from the given μ), just as it is for the closed system (where a fixed N can be used to define a derived μ when necessary).
2. We have seen above that for the open MB gas, the distribution is $\overline{n_i} = \gamma_i = \exp[(\mu - \varepsilon_i)/k_B T]$. Replacing μ from (13.14), we see that this can be written as

$$\overline{n_i} = \frac{N}{Z}\exp(-\varepsilon_i/k_B T)$$

the familiar form for the MB distribution, again with the subtle difference that N is now a derived not a given quantity. As expected all this is consistent with our earlier discussion of the multiplier α and its relation to chemical potential, as in (13.1).

13.4 MIXED SYSTEMS AND CHEMICAL REACTIONS

Having discussed open one-component assemblies, we are now ready to consider what happens when more than one substance is present. This will then enable us to determine the equilibrium conditions for simple chemical reactions. We shall find that the chemical potential plays a crucial role in understanding chemical reactions (not too surprisingly when you think of the name!).

13.4.1 Free energy of a many-component assembly

In section 13.1, (equation (13.2)), we saw that, in a one-component assembly, we could write the chemical potential μ as the Gibbs free energy per particle, so that $G = N\mu$.

The generalization to a many-component assembly is straight-forward from this standpoint. It is that each species (component), labelled by s in what follows, has its own chemical potential μ_s which is again the Gibbs free energy per particle of species s. Hence we may always write for the total Gibbs free energy of the assembly

$$G = \sum_s N_s \mu_s \tag{13.15}$$

The sum goes over all components s of the assembly (e.g. s = nitrogen, oxygen, carbon dioxide, water, ...if we were considering air), and there are N_s particles of component s. It is worth stressing that (13.15) is of very general applicability, although we shall only illustrate its use in gaseous assemblies in what follows.

13.4.2 Mixed ideal gases

The description of mixed ideal MB gases follows naturally from the idea of chemical potentials and from the discussions of section 13.3 (especially section 13.3.3). The grand canonical ensemble now relates to an assembly of given volume V, given temperature T and given chemical potentials μ_s for each and every component s. Since we consider only ideal gases, there is no interaction between the various components. Therefore the energy levels for each component are the same as they would be by themselves in the ensemble. Hence the grand partition function Z_G factorizes as follows:

$$Z_G = \prod_s Z_G^{(S)} = \prod_s \left[\prod_j \exp(\gamma_j^{(s)}) \right] \tag{13.16}$$

In the first part of (13.16), $Z_G^{(S)}$ represents the contributory factor to Z_G from component s, as mentioned above. The last part comes about when we introduce the requirement that each component is a gas in the MB limit. The factors $\gamma_j^{(s)}$ for each component are defined exactly as in (13.7) with energy values $\varepsilon_j^{(s)}$ appropriate to the gas in question.

As in section 13.3.3, we now derive thermodynamic quantities such as P and the numbers N_s of each component. Equation (13.5) for PV tells us that

$$\frac{PV}{k_B T} = \ln Z_G = \sum_s \ln Z_G^{(s)} = \sum_s \exp(\mu_s/k_B T) Z^{(s)} \tag{13.17}$$

where $Z^{(s)}$ is defined as the one-particle (ordinary) partition function for component s (compare (13.13) for the one-component equation).

The numbers of each gas component are derived from the many-component generalization of (13.5) and (13.6), following (13.15)

$$N_s = \left(\frac{\partial (PV)}{\partial \mu_s} \right)_{V,T,\mu(\text{other components})}$$

Substituting PV from (13.17) gives

$$N_s = k_B T \frac{\partial}{\partial \mu_s}(\ln Z_G) = \exp(\mu_s k_B T) Z^{(s)} \tag{13.18}$$

Comparing (13.17) and (13.18) shows that $\ln Z_G = \sum_s N_s$ and hence

$$\frac{PV}{k_B T} = \sum_s N_s$$

is the appropriate generalization of the equation of state. The same result can be written as the 'law of partial pressures'

$$P = \sum_s P_s = \sum_s (N_s k_B T / V) \tag{13.19}$$

i.e. the total pressure P of the ideal gas mixture is the sum of the pressures P_s which each component would exert in the absence of the others.

13.4.3 Equilibrium in chemical reactions

In section 13.4.4 we shall consider a reaction in the gaseous phase, so that the treatment of the last section remains relevant. However the memorable result of this section requires no such restriction. To be specific, let us consider a reaction of the type

$$\mathrm{A} + \mathrm{B} \rightleftharpoons \mathrm{AB}$$

in which an atom of component A combines with an atom of component B to give a molecule of component AB. (An apology. This is not such a common reaction type, especially in the gaseous phase, although it is the simplest. More familiar are reactions such as $H + H \rightleftharpoons H_2$, or $H_2 + Cl_2 \rightleftharpoons 2HCl$, or $2H_2 + O_2 \rightleftharpoons 2H_2O$.

These can each be treated by methods which are entirely similar in principle, but which have some differences in detail from our chosen simple reaction. For instance the first has two identical atoms on the left, and the other two reactions similarly have 2s appearing.)

For a gaseous phase reaction, the results of the previous section, such as the law of partial pressures, are all entirely valid, where we identify s with A, B and AB. But now there is another requirement, that of chemical equilibrium, to consider. We can write the free energy (equation (13.15)) as

$$G = N_A \mu_A + N_B \mu_B + N_{AB} \mu_{AB} \tag{13.20}$$

Equilibrium means, as ever, that we should minimize the free energy (see section 13.1 above, for example). In the grand canonical ensemble approach, the

μs are fixed, so that the requirement is that (13.20) must be stationary if we make small changes in the numbers of reacting molecules. We must satisfy the equation

$$dG = \mu_A dN_A + \mu_B \, dN_B + \mu_{AB} \, dN_{AB} = 0 \tag{13.21}$$

However, such changes of number are not independent. If we increase the AB population by one, we are of necessity reducing both the A population by one and the B population by one, as can be seen from the reaction equation. (Now does it become clear how to generalize the approach to other reaction types?) In other words, the allowable variations are limited to $dN_{AB} = -dN_A = -dN_B$. Substitution of this restriction into (13.21) immediately leads to the conclusion that in equilibrium we must have

$$\mu_A + \mu_B = \mu_{AB} \tag{13.22}$$

This is a famous (and simple) result. It tells us that when chemical equilibrium is reached, the chemical potentials of the reactants must satisfy a relationship of this type. They are no longer independent.

13.4.4 The law of mass action

When equilibrium in our reaction is reached, we know (from (13.22)) that $\mu_A + \mu_B = \mu_{AB}$. Hence

$$\exp(\mu_A / k_B T) \cdot \exp(\mu_B / k_B T) = \exp(\mu_{AB}) / k_B T \tag{13.23}$$

Now we return to gases. We now assume that A, B and AB are all ideal MB gases, so that we can use the results of section 13.4.2. Equation (13.18) tells us how the number of each component s is related to its chemical potential and its partition function $Z^{(s)}$. This can be used to derive a relation between the numbers of reacting components in chemical equilibrium. In fact, using (13.18), (13.23) can be written simply as

$$\frac{N_A}{Z^{(A)}} \cdot \frac{N_B}{Z^{(B)}} = \frac{N_{AB}}{Z^{(AB)}}$$

which can be rearranged to read

$$\frac{N_{AB}}{N_A N_B} = \frac{Z^{(AB)}}{Z^{(A)} Z^{(B)}} = K(V, T), \text{ say} \tag{13.24}$$

This is the 'law of mass action' for the reaction $A + B \rightleftharpoons AB$. It relates the number of each component to a so-called equilibrium constant $K(V, T)$ which depends only on the reaction, the volume and the temperature. Similar laws of mass action can be developed for other reactions using the same methods. Once $K(V, T)$ is known, the law enables us to predict what will be the equilibrium composition of the reacting mixture from any given starting conditions.

There is an important word of warning in the calculation of $K(V,T)$. The partition functions for the various components must all be calculated using the same energy zero. (That is where a lot of the chemistry comes in. After all, you expect the reaction to go from a high energy state to a low one!) What this means is that the partition functions take the form

$$Z^{(s)} = V \left(\frac{2\pi M^{(s)} k_B T}{h^2} \right)^{3/2} \cdot Z_{int} \cdot \exp(-W^{(s)}/k_B T)$$

where in the final term a constant $W^{(s)}$ is included to get the energy levels of that component right, i.e. to get the ground state of the translational motion (usually taken as zero in a one-component situation) at the right energy. The result of this is that $K(V,T)$ will contain a factor of the form $\exp(-\Delta W/k_B T)$, where in this case $\Delta W = W^{(AB)} - W^{(A)} - W^{(B)}$. The sign and magnitude of ΔW has a great influence on the equilibrium condition of the reaction. If ΔW is positive, the forward reaction will be a reluctant starter particularly at low temperatures, since the exponential factor in K will be small and hence N_{AB} will also be small. On the other hand, a negative ΔW will favour the forward equilibrium of the reaction, particularly at low temperatures. The exponential will be large and hence so will be its effect on K and on N_{AB}. This is as expected. A negative ΔW means that the AB side of the equation has a lower energy than the A + B side.

Equation (13.24) also allows us to understand the volume dependence of the reaction. All the partition functions have V as a simple factor arising only from the translational part. Hence $K \propto 1/V$. Thus a small volume will favour the forward reaction. Again this is reasonable, since the formation of AB causes a reduction in the total number of molecules, hence a reduction in the pressure (equation (13.19)) and hence in the PV energy of the system.

13.4.5 Reaction rates

The law of mass action (13.24) tells us the equilibrium concentrations in a reaction. That is a fine achievement, but it is not always enough to do justice to chemical reality. For instance, when we consider the reaction A + B → AB, we have seen above that when ΔW is negative the yield of AB will tend to improve at lower temperatures. However, as any cook knows, reaction *rates* get smaller at lower temperatures, so that one may have to wait an unrealistic time for the theoretical equilibrium to occur. The reaction can be frozen into a metastable state in which the chemical potentials do not have time to equalize. This is a familiar thermodynamic situation. We have discussed earlier (see Chapter 10) the existence of frozen-in disorder (i.e. non-zero entropy) at low temperatures.

In chemical reactions, the usual situation is that illustrated in Fig. 13.1. Consider the transition between separate A and B atoms towards the formation of an AB molecule, for a system in which ΔW is negative. The figure shows a schematic graph showing

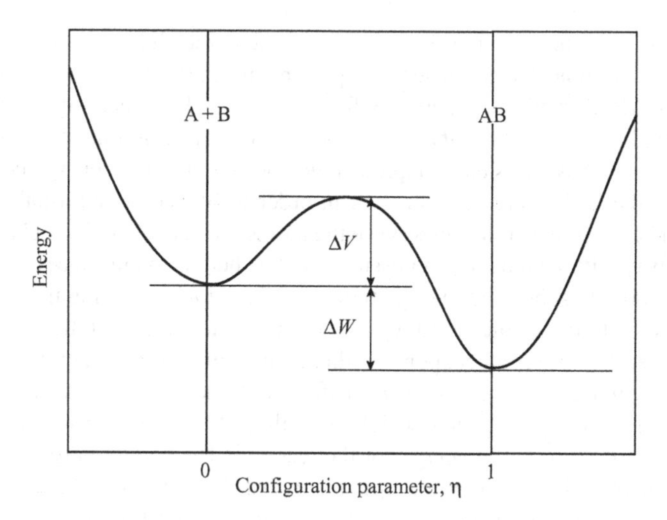

Fig. 13.1 Schematic graph of the energetics of a chemical reaction. The valley at $\eta = 0$ represents separate A and B atoms. The valley at $\eta = 1$ represents the AB molecule.

the form of the free energy plotted against a 'configuration parameter', η say. We have defined $\eta = 0$ to correspond to the system when it consists of A + B, and $\eta = 1$ when the system is AB. We note that the AB valley at $\eta = 1$ is lower than the A + B valley at $\eta = 0$, the difference in height being effectively the energy ΔW discussed above. However for the reaction to occur, the pair of A + B atoms must somehow deal with the barrier of height ΔV which exists in the transition process at intermediate values of η.

There is a very pretty analogy here. Consider Fig. 13.1 as a description of a mass of water in the Earth's gravitational field, i.e. it can be thought of as a graph of gravitational potential energy (essentially height above sea level) against distance. The lowest energy state is in the deep valley at $\eta = 1$. The gravitational potential energy of the water which starts in the high valley at $\eta = 0$ is higher, so that energy ΔW is available if the transition can be made (ask any hydroelectric engineer). But in order to effect this transition, one has to do one of three things: (i) tunnel through the hill, a frequent ploy for hydroelectric power, or (ii) if it exists, find some other way around which is not shown in the section shown in the figure (perhaps they are the same valley curved around in another dimension?), or (iii) pump or carry the water over the hill, requiring at best the loan of an amount of energy ΔV.

The same thinking helps us with the chemical reaction. What can make the reaction go? There are three possible routes. Route (i) involves tunnelling through the barrier. Although for a microscopic system this is in principle possible by quantum tunnelling through the barrier, for most ordinary chemical reactions this has such a low probability that it is entirely negligible. Route (ii) involves evading the barrier, and this is a way of cheating commonly used in chemistry. It is the principle of using a catalyst which allows another path not shown on the graph, which has a lower barrier or none at all. As an example, let us consider one possible mechanism for the gaseous reaction $A + B \rightarrow AB$. Perhaps the reaction will only proceed if we succeed in getting A and B sufficiently close together that an electron can jump from one atom to the other. In the gaseous phase, this close approach may need a very high energy collision, the reason for the high energy barrier ΔV, in order to overcome the usual short-range repulsion between the atoms; otherwise the atoms will merely collide elastically and bounce away without reaction. A possible catalyst here is to introduce a large surface area of a suitable substance (e.g. platinum black, very fine platinum metal). Atoms can now stick to the surface for a while and, given the right catalytic material, come into sufficiently close contact upon it, rather than having to await a mid-air encounter.

However, in the absence of such an artificial aid, the only route remaining is route (iii), finding enough energy to get right over the ΔV barrier. For this reason ΔV is often termed the *activation energy* for the reaction. The energy must be found from the high energy tail of the thermal Boltzmann distribution. The collision mechanism mentioned above can be used as an illustration of what this implies. A collision between an A and a B atom will usually not cause a reaction. Only in the (unlikely) event of a near head-on collision between two fast-moving atoms will the close encounter take place. This implies a reaction rate R (reaction probability per second) of the form $R \simeq f \exp(-\Delta V/k_B T)$, where f is an attempt frequency (the collision rate) and the exponential Boltzmann factor represents the probability of the energy condition being satisfied. Needless to say, an exact treatment of reaction rates is much more elaborate than this quick sketch. However it gives the correct flavour for almost all reactions in solids, liquids or gases. Thermal activation over an energy barrier always invariably plays a fundamental role and its probability is well governed by the Boltzmann factor.

In many chemical reactions, the activation energy is of order of 0.5–1 eV, the typical energy scale for any electronic transitions in atoms. Thus it is that the Boltzmann factor can be very small (say 10^{-10} at room temperature), and hence also the temperature dependence of R is very fierce. A small temperature rise often gives a marked increase in rate, as a cook's timetable shows; and also the rate R becomes effectively zero when the temperature falls significantly, leading to metastability (i.e. zero reactivity in this context) as stated at the start of this section.

14

Dealing with interactions

So far in this book we have dealt almost entirely with assemblies made up of weakly interacting particles, either localized or gaseous. This has really not been a matter of choice, but almost of necessity. The basic approach has been founded on the idea that the energy of the assembly is simply the sum of the individual one-particle energies, i.e: $U = \sum n_j\varepsilon_j$. Without this simplification the mathematics gets out of hand (although some progress can be made if we start with the canonical or grand canonical approach).

Nevertheless we have successfully applied our statistical ideas to a rather wide variety of real problems. In the case of real chemical gases, this success is not too much of a surprise, since it has been well known since the days of kinetic theory that gases are almost ideal, at any rate at low pressures and not too close to the liquid phase transition. Later in the chapter, we shall discuss briefly how the small corrections to the ideal gas equation may be calculated, to give a more realistic equation of state than $PV = Nk_BT$.

But chemical gases are not the only situation in which ideal statistics are applied. There are several cases where the assumption of weakly interacting particles must cause raised eyebrows, in spite of our earlier protestations. These include (i) treatment as a Fermi gas of the conduction electrons in metals and semiconductors (section 8.2), (ii) treatment of liquid helium-3 as an ideal FD gas (section 8.3) and (iii) treatment of liquid helium-4 as an ideal BE gas (section 9.2). We take the opportunity in this chapter to explain a little further our understanding about why the simple models turn out to be applicable.

The central idea to develop is that of *quasiparticles*. We concentrate not on the 'real' particle, but on some other entity called a quasiparticle. The quasiparticle is chosen so that it is weakly interacting to a better approximation than is the original particle. One way in which we have already seen this type of approach at work is in our treatment of lattice vibrations of a solid. The motions of the atoms themselves are very strongly interacting in a solid, so instead we choose to redefine the total motion in terms of phonons (see section 9.3.2), which can be treated as a weakly interacting ideal gas model to a fair approximation. Let us now see how this works out for the three cases listed above.

14.1 ELECTRONS IN METALS

The elementary treatment of electrons in metals (section 8.2) is simply to describe the electrons as an ideal FD gas. This was first done in the early days of quantum theory (the Sommerfeld model), and it is surprisingly successful in describing the results of experiments (on both equilibrium properties and transport properties), particularly if parameters like the number density and the mass of the electrons are taken as adjustable parameters. In the case of the simplest metals, such as the alkali (group 1, monovalent) metals sodium and potassium, these adjusted parameters are in fact very similar to the free electron parameters. So are they (significantly) for most liquid metals. However they become quite fanciful for some other crystalline solids, even simple elemental ones, such as the group 4 semiconductors germanium or silicon, not to mention the insulator carbon (diamond), in which the effective conduction electron density is clearly zero.

Although the whole basis of the Sommerfeld model was long known to be questionable, it nevertheless took scientists at least 30 years to begin to understand why the model works so well. The problem is obvious. The Fermi energy which we calculated in Chapter 8 for a typical simple metal is of order 5 eV. This is the kinetic energy scale for the supposed free electrons, and it is large enough compared to a thermal energy scale $k_B T$ at any reasonable temperature to make the supposed electron gas an extremely degenerate FD gas, as we have seen. However, this energy is not at all large compared with the potential energy of interactions which we should expect between an electron and its surroundings. There are two problems here. Firstly, there is the interaction between a conduction electron and the lattice ions. The conduction electrons are negatively charged, having been liberated from atoms leaving behind positively charged ions. Therefore there will be a strong attractive Coulomb interaction between them, of expected magnitude about $e^2/4\pi\varepsilon_0 r$ where r is about an atomic spacing. Putting in $r = 0.2$ nm, this gives an energy of 7 eV, of the same order of magnitude as the kinetic energy. Secondly, there is also the potential energy of interaction between the conduction electrons themselves. They are not truly weakly interacting, since they repel each other with a repulsive interaction which again should be of the same order of magnitude, since the electrons are typically an interatomic spacing apart.

Where does this leave us? Clearly in a great mess, since this last interaction in particular blows away the whole 'weakly interacting' assumption behind the ideal FD gas treatment of the electrons. However, the reason why all is not lost is not hard to see at the hand-waving level. Overall, the metal is electrically neutral. Therefore problem 1 (electron–ion attraction) and problem 2 (electron–electron repulsion) must to a good approximation cancel out. This is certainly true of the long-range part of these interactions, and this is the major justification for continuing with the simple approach. The idea is that of screening (electrostatic shielding). The 'other' electrons cancel out the effects of the positive lattice ions.

Interestingly, this is not yet the end of the story. We are still left with what would be a very substantial short-range lattice potential, and a residual worry about electron–electron interactions. In the 1960s it was realized that the problem of the large lattice potential could be attacked in a rather clever imaginative way, all as a consequence of the Pauli exclusion principle. This is where the quasiparticle ideas come in, although the language was not applied in context at the time. The point is the ion itself is not a simple mass with a positive charge. In sodium, for instance, it consists of a nucleus with 11 protons, surrounded by 10 closely bound ('core') electrons, filling the 1s, 2s and 2p orbitals. The one conduction electron per atom was given to the gas-like pool (the so-called Fermi sea) from the outer (partially filled) 3s shell of the isolated atom. Now the Pauli principle tells us that these conduction electrons in the metal must of course be 'unlike' the bound core electrons in each atom. In technical language, they must be in states which are orthogonal to the core electron states. In practice, this means that they have wave functions very much like the 3s functions when they are near an ion core. Now here is the clever part. The 3s wave function is a very wiggly affair near the nucleus – and our conduction electron is forced into these wiggles by the Pauli requirement for it to be dissimilar from the filled core states. And wiggles in wave functions mean high curvature ($\partial^2\psi/\partial x^2$ etc.) and hence high kinetic energy. So the wave function is made to contribute a high kinetic energy in just those regions of space close to the nucleus where we know the potential energy must be large and attractive. There is a pretty cancellation here. This approach is called a pseudo-potential approach. The detailed mathematics and justification of it is well beyond the scope of a small book on statistical physics. But the idea is important. What we do is to convince ourselves that the real problem (of a true conduction electron, moving through a large albeit short-range attractive lattice potential) has considerable similarity to, and in particular the same one-particle energies as, a pseudo-problem (of an essentially free electron moving through a much smaller lattice potential). The pseudo problem is solvable, as in the 1930 treatments, by perturbation from the free electron model with the perturbing potential now a small quantity which can be of either sign. It becomes an adjustable parameter (with the adjustments made by experimentalists or by theorists, who on a good day and for a simple metal come up with the same answers).

Although this to an extent justifies the FD gas approach, we are still left with the residual effects of our two potential energy problems.

First, let us continue to ignore any effects of electron–electron interactions. In other words, we adopt a one-electron approach which is tractable according to an ideal gas model. The basic assumption that there are one-particle states in which to set the problem remains; we may write $U = \sum n_j\varepsilon_j$, and we may enumerate the states by fitting waves into boxes as in Chapter 4. What differs from the free electron gas, however, turns out to be very profound. It is in the dispersion relationship, needed to convert k (from the waves in the box) into energy ε (to use the FD distribution). Even a small periodic perturbation, such as that from a crystalline (pseudo-) lattice, has a dramatic effect on the states near a Brillouin zone boundary, where k for the electron has a relationship with the lattice periodicity. The result is that the Fermi surface (see Fig. 8.2) no longer remains a simple sphere (except almost for the monovalent alkali metals) but becomes carved up and somewhat distorted or 'sandpapered' by the lattice perturbation. The model is no longer isotropic, but the $\varepsilon - k$ relation can be markedly different in different crystallographic directions. So it is that the parameters of the model can be changed dramatically, particularly in polyvalent metals such as aluminium. The original Fermi sphere is chopped up into a number of pieces by the pathological $\varepsilon - k$ relation, notably the existence of energy gaps at the Brillouin zone boundaries. In many metals or in semimetals like bismuth, the Fermi surface then contains small pieces, which can have electron-like or hole-like character; for instance, a dispersion relation over the relevant range of the form $\varepsilon = A + \hbar^2 k^2/2m_1$ would correspond to electrons of effective mass m_1, whereas $\varepsilon = B - \hbar^2 k^2/2m_2$ would correspond to holes (because of the minus sign) of effective mass m_2. But it is important to stress that the whole statistical treatment remains valid. There is still a Fermi surface, and the thermal properties are governed precisely by the density of states $g(\mu)$ at the Fermi level. The complication comes merely in the calculation of $g(\varepsilon)$.

This one-electron approach has been the basis of much successful modelling of the thermal and transport properties of real metals. But it is still only an approximation because of the continuing neglect of what are called many-body interactions. These are still in fact quite significant in many metals, particularly in the transition metals. They can be observed experimentally in several ways, of which the following is one:

1. Measure the electronic contribution to the heat capacity, from low-temperature experiments. This is identified as the linear term γT in the experimental $C = \gamma T + \beta T^3$; the T^3 term is the phonon contribution (Chapter 9), whilst the linear term is the required electronic contribution.
2. Calculate from the $\varepsilon - k$ curves the electron density of states $g(\mu)$ at the Fermi level; in practice this can often be done rather accurately, since there are many checks in the calculation, such as information from specialist Fermi surface measurements. Hence calculate a so-called 'band structure' heat capacity γ using this density of states (see (8.11)).
3. Note that the measured value is (often) bigger than the calculated one by an enhancement factor (η, say) which is greater than 1, and in practice can be as large as 2 or 3. For instance η is about 1.2 even in the ideal sodium, it is 1.5 in aluminium, about 2.1 in lead and even larger in the transition metal palladium.

There are (at least) three reasons known for such many-body effects, namely electron–electron interactions, electron–phonon interactions and interactions involving magnetic excitations. We have mentioned electron–electron interactions above. But actually they are quite small (except in some transition metals, also the only candidates for the magnetic enhancement). The reason for this is found in the Pauli principle, which tends to keep electrons apart, quite apart from the Coulomb repulsion. Hence only the long-range part is relevant and this has been used up in screening the lattice ions (above).

The major effect in simple metals comes from the electron–phonon interaction. One electron passing through the lattice can excite a short-lived phonon which is then sensed (absorbed!) by a second electron. This rather subtle interaction is responsible for superconductivity, to which we shall return briefly in section 15.1.1. It is also responsible for most of the enhancement factor. The effect is to couple together electrons which can scatter (emit or absorb) phonons, i.e. those whose energies lie within about $k_B\theta_D$ of the Fermi energy. (The Debye temperature θ_D is a scale temperature for phonons, typically about room temperature for many solids, see section 9.3.2. It is much less than the Fermi temperature for a metal.)

The result of this scattering, or mixing together, of one-electron states is illustrated in Fig. 14.1. It is the one-electron states close to the Fermi energy (just the ones we are interested in!) which become mixed up with each other by the interaction with the phonons. The result is that the effective density of states at μ is *increased* above

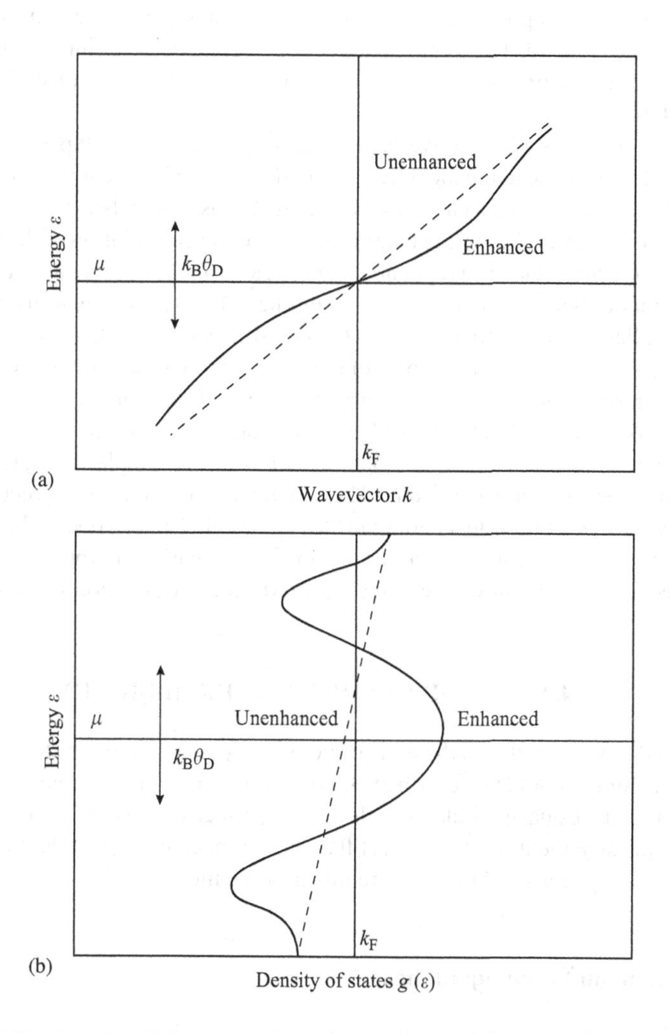

Fig. 14.1 (a) The dispersion relation and (b) the density of states for a free electron gas (dashed curves) and for an electron gas with interactions (full curves, 'enhanced').

the one-electron value at the expense of other states within a few $k_B\theta_D$ of the Fermi level, as shown in Fig. 14.1b. Incidentally, this figure also demonstrates why it is that at high enough temperatures ($T > \theta_D$) the enhancement is shed, i.e. $\eta = 1$. This occurs because the experimental 'sampling window' of the $g(\varepsilon)$ curve is the thermal energy scale $k_B T$. When this window exceeds $k_B\theta_D$, the whole of the deformed curve is sampled rather than simply $g(\mu)$. And since the interactions do not generate more states, but simply move the energies of the old one-electron states, the average density of states is unchanged. This shedding effect is observed experimentally. It is a fine challenge to relate the dispersion curve (Fig. 14.1a) to Fig. 14.1b. It shows the same physics! The states are evenly spaced in k, as ever. The flattening of the $\varepsilon - k$ curve at the Fermi level implies that there are *more* k-states per unit energy range than before, as stated above. The joining back to the original curve ensures a higher slope at energies a little removed from μ and hence a diminished density of states as shown in Fig. 14.1b.

To summarize this section. We have seen that there are two types of correction which need to be made to the ideal free gas model in describing conduction electrons. Fortunately, both effects are capable of being treated as 'small' because (i) the overall electrical neutrality of the metal guarantees an accurate cancellation of the long range parts of the electron–electron repulsion and the electron–lattice ion attraction and (ii) the high kinetic energy of conduction electrons near the ion core allows us to deal with a much smaller *effective* lattice potential than the true potential. The first consideration is the effect on the free electron model of the small effective electron–lattice potential, namely to introduce the lattice symmetry into the band structure or $\varepsilon - k$ relation for the electrons. As a result, although FD statistics are still valid, they must be applied with the new and intricate $\varepsilon - k$ relation, not with the simple free gas one. This is the 'one-electron approximation'. The second consideration is to recognize that there may well be other effects going on which are of a 'many-body' character. The residual electron–electron interactions are an obvious candidate for such effects, but in practice the largest many-body effects arise from electron–phonon interactions.

14.2 LIQUID HELIUM-3: A FERMI LIQUID

In section 8.3, we introduced the topic of liquid helium-3. The suggestion made there was that the liquid could be described as a free fermion gas, that the gas is not of bare ^{3}He atoms but of quasiparticles which have an effective mass of several bare ^{3}He masses. This statement has the correct flavour, but needless to say the full truth is a little more complicated. Let us explore this in some more detail.

14.2.1 Landau Fermi liquid theory

As stated in section 8.3, our understanding of so-called Fermi liquids owes much to the Russian theorist, Lev Landau. A fuller treatment of this section may be found in specialist low-temperature physics books under the banner of *Landau Fermi liquid theory*. What Landau did was to show that for many purposes the energy of an assembly of N ^{3}He atoms in the liquid can indeed be written as

$$U = \sum_j n_j \varepsilon_j \qquad (14.1)$$

where $\sum n_j = N$ as usual. This looks like an ideal gas assumption. However, the appropriate energies ε_j are no longer the free-particle energies. Instead they are some quasiparticle energies which allow for the fact that when you touch one ^{3}He particle you touch them all! Anyone who has stirred his tea knows that this is bound to be the case in a liquid. If you move a particular atom from A to B in a liquid, then there are consequences for the 'other' atoms also. The liquid very strongly wants to maintain a uniform density, so that there is clearly a 'backflow' contribution to the motion – you need to move some background fluid out of the way at B and into the hole at A. There is also the fact that you cannot get hold of one atom without involving the interactions with its neighbours. So it is that a first approximation is to talk about quasiparticles with an effective mass higher than the bare mass.

The problem is, of course, truly a many-body problem, similar to that discussed in the previous section. The quasiparticle energies ε, being related to the other particles present, themselves must depend on the distribution of these other particles. Furthermore, the influence of these other particles will depend on the property being discussed. For example, above we discussed a simple A to B motion, but what about reversing the spin of a particle or passing a sound wave through the liquid? Thus it is not surprising that the liquid is not describable simply by a one-parameter correction such as a single effective mass. Rather a range of parameters is required. The clever part of Landau's work was to show that only a few such parameters are needed in practice, when you are considering a Fermi liquid at low enough temperatures.

In outline, the treatment works as follows. The essential idea is to convince oneself (not obvious, unless you are called Landau!) that the only significant effect of the interactions is to *change the dispersion relation* for the quasiparticles. All the fitting waves into boxes (Chapter 4) still works, so that the definition of k-states and k-space and the density of states $g(k)$ is all unchanged from the ideal gas. All that happens is that the energy ε of a quasiparticle is changed from the ideal gas. All that happens is that the energy ε of a quasiparticle is changed from the ideal gas $\varepsilon = \hbar^2k^2/2M$. Also unchanged are the use of the number condition to determine the Fermi energy μ and the use of the FD distribution (8.2) to give the thermal equilibrium occupation numbers of the quasiparticle states.

So, how is ε now worked out? The idea is really rather pretty. You concentrate only on energy changes (and, after all, that is all that any experiment can determine). Landau considered the change δU in the internal energy of the whole assembly when a change $\delta f(\boldsymbol{k})$ is made in the distribution function $f(\boldsymbol{k})$. In this treatment, we use the vector $\boldsymbol{k}$ as a label for the k-states as introduced in (4.2), and we use the notation $\delta\boldsymbol{k}$

for a volume element in k-space. Landau's approach then writes

$$\delta U = \int g(\boldsymbol{k})\varepsilon\delta f(\boldsymbol{k})\mathrm{d}\boldsymbol{k} \tag{14.2}$$

an equation looking very like the non-interacting $U = \sum g_i \varepsilon_i f_i$ (compare (14.1)) but one which is now used to *define* the excitation (quasiparticle) energy ε.

In the Fermi quasiparticle gas at low temperatures, only excitations close to the Fermi surface are important, so we may write (to first order)

$$\varepsilon(k) = \mu + \left(\frac{\partial \varepsilon}{\partial k}\right)_{\mathrm{F}} (k - k_{\mathrm{F}})$$

The usual group velocity definition of the Fermi velocity gives us quasiparticle velocity

$$v_F = \frac{1}{\hbar}\left(\frac{\partial \varepsilon}{\partial k}\right)_{\mathrm{F}}$$

and we may also define an effective mass from the momentum relation $M^* v_{\mathrm{F}} = \hbar k_{\mathrm{F}}$. It is worth stressing that these values of v_{F} and of the effective mass will be different from the free gas values, because of the new definition of ε. In addition, we can show that the density of states at the Fermi level $g(\mu)$ is changed; it is identical in form to that for the ideal gas (8.3), but now the mass M must be replaced by M^*. (This is a result worth proving from the above, as an exercise.)

The final illumination of the Landau theory is to separate out the effect of the interactions on these quasiparticle energies. Again, the theory concentrates on energy *changes*. The idea now is to look at the energy shift of a particular quasiparticle state labelled by $\boldsymbol{k}$. Its energy in equilibrium at $T = 0$ is $\varepsilon_0(\boldsymbol{k})$, and we ask what energy shift takes place when we change the occupation numbers of all the other states (labelled by $\boldsymbol{k}'$). The answer is

$$\varepsilon(\boldsymbol{k}) - \varepsilon_0(\boldsymbol{k}) = \int \frac{F^s(\boldsymbol{k}, \boldsymbol{k}')}{g(\mu)} \delta f(\boldsymbol{k}') g(\boldsymbol{k}')\mathrm{d}\boldsymbol{k}' \tag{14.3}$$

The 'interaction function' $F^s(\boldsymbol{k}, \boldsymbol{k}')$ is defined in this way because (i) it is dimensionless and (ii) it is zero in the absence of interactions. For our low-temperature Fermi liquid, it is obvious that changes in occupations of states $\boldsymbol{k}'$ which are near to the Fermi surface are the only ones of any relevance to measured properties. Since all k values are thus essentially equal to k_{F}, a further simplification may be made, namely that the interaction function F^s may be considered as a function only of the angle, θ say, between $\boldsymbol{k}'$ and $\boldsymbol{k}$. The honest person's mathematical treatment is then to expand $F^s(\theta)$ in terms of the suitably orthogonal 'Legendre polynomials' to give

$$F^s(\theta) = F_0^s + F_1^s \cos\theta + F_2^s (3\cos^2\theta - 1)/2 + \cdots$$

However, since most measured properties involve rather smooth functions of angle for the relevant $\delta f(\boldsymbol{k})$ variations, it is no real surprise that most of the physics is found in the first two or three of these F^s numbers. The numbers are often called Landau parameters.

Actually, we have over-simplified the situation in one important respect. As stressed in Chapter 4, 'quantum states are k-states plus'. The interaction function must also specify the spins of states $\boldsymbol{k}'$ and $\boldsymbol{k}$. Hence, besides the space-dependent F^s parameters there is also a spin-dependent set of corresponding F^a parameters, F_0^a, F_1^a etc. These will be important when a magnetic field is applied, to give a spin-dependent $\delta f(\boldsymbol{k})$. The usual superscripts stand for symmetric s and antisymmetric a.

14.2.2 Application to liquid ^{3}He

When we interpret experimental results using Landau theory, we find that (i) the corrections to ideal gas theory are really large, but (ii) the theory works remarkably well.

Most properties are explained using only the three parameters F_0^s, F_1^s and F_0^a. For instance, the effective mass turns out to depend on F_1^s (the $\cos\theta$ treats north and south differently, as required for A to B motion), and calculation shows that $M^*/M = 1+F_1^s/3$. The effective mass is experimentally accessible through the density of states, i.e. from the heat capacity, which is proportional to M^*. As an example, consider pure liquid ^{3}He at low temperatures and at zero pressure – remember it can be studied at pressures up to 35 bar before it solidifies. As noted in Chapter 8, the heat capacity is found to be linear in the millikelvin range, characteristic of an ideal FD gas. However, the magnitude of C at zero pressure is about 2.76 times larger than expected from ideal gas theory with the normal ^{3}He mass. This is explained in Landau theory by setting $2.76 = M^*/M = 1 + F_1^s/3$, and hence $F_1^s = 5.3$. As noted above, $F = 0$ would imply no interactions; this is a big effect, and it gets even bigger as pressure is applied, reaching $F_1^s = 14.6$ at the melting pressure (35 bar).

The first Landau parameter F_0^s turns out to be even larger ($F_0^s = 9.2$) for ^{3}He at zero pressure. This interaction term is uniform over all states, so it comes into play when the density changes as in a sound wave. The value of 9.2 derives from the experimental sound velocity of around $180\,\mathrm{m\,s^{-1}}$.

The value of F_0^a is obtained from measurements of the magnetic susceptibility (see (8.13)). This depends on the density of states (as does the heat capacity) but also on the tendency for alignment of the spins; the discussion of ferromagnetism in section 11.2 is relevant. In fact the susceptibility is about a factor of 3 higher than would be expected from non-interacting spin-$\frac{1}{2}$ fermions in a gas with the density of states determined from the heat capacity. Spin-dependent interactions enter through a factor $(1 + F_0^a)^{-1}$ in the susceptibility, and the experimentally determined value of F_0^a is – 0.7. This is a very strong coupling, with the negative sign implying that interactions tend to align the spins. It is worthy of note that a value for F_0^a of -1 would imply that the liquid was ferromagnetic!

The Landau parameters relate to other measured quantities also, for instance the viscosity and diffusion coefficients in the liquid. These transport properties can be explained on Fermi liquid theory with a remarkable efficiency and using no new assumptions. The only additional idea to be introduced (and worked out using the theory) is that of the mean free path l for ^{3}He–^{3}He scattering. Because of the Exclusion Principle, scattering can only take place for quasiparticles with energies up to about $k_B T$ from the Fermi energy. This leads to the result $l \propto T^{-2}$ in the degenerate FD regime, so that the mean free path, and with it the viscosity of the liquid, gets extremely large at low temperatures. For example, the viscosity of pure ^{3}He just above its superfluid transition at 1 mK is about the same as the engine oil in a car! (See Fig. 15.3 for some measurements related to this viscosity; the fluid thins out dramatically again when it becomes a superfluid.)

Therefore we now have a very full understanding of the properties of liquid ^{3}He. If that were the end of the Landau story, it would be clever, but not world-shattering. However, there is even more to relate. The really splendid thing is that Landau's theory predicted some entirely new effects. The most important of these is called *zero sound*. Normal sound in a gas does not propagate when the mean free path becomes large enough; a bell cannot be heard through a vacuum. The density gradients of the sound wave cannot be established when the scattering length becomes larger than the sound wavelength. However in the interacting Fermi liquid, an entirely new type of collisionless collective motion of the liquid becomes possible, with the restoring force deriving from the interactions. Thus as the temperature is lowered (to increase l) or as the sound frequency is raised (to lower the wavelength), it is observed that the sound velocity shifts (from around 180 to 190 m s^{-1}) and there is also a large attenuation in the changeover regime, as normal ('first') sound

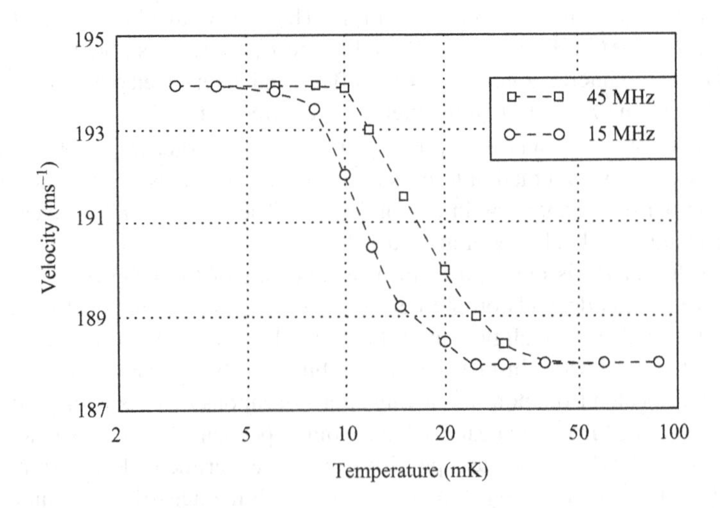

Fig. 14.2 The velocity of sound in liquid ^{3}He at millikelvin temperatures. The measurements show the transition from first (normal) sound at high temperatures to zero sound at low temperatures.

changes to the new mode of zero sound. Some experimental results showing this effect are shown in Fig. 14.2. The rich experimental data here is entirely as predicted by Landau theory, and gives a striking confirmation of the validity of the whole approach.

Finally in this section, we can remark that there are many forms of Fermi liquid available to the ^{3}He physicist and for which Landau theory gives a good description. Besides pure ^{3}He as a function of pressure (and hence density), there is also the whole topic of ^{3}He–^{4}He solutions. At low temperatures (say, below 100 mK), these solutions have remarkably simple properties. The ^{4}He component of the solution is thermally dead. It is well below the superfluid transition temperature, so that it has already reached its zero entropy ground state (see section 9.2, and in particular Fig. 9.6). Thus all of the thermal action belongs to the ^{3}He component only. Nature is kind here, in that there is a substantial solubility range of ^{3}He in ^{4}He, from zero up to a 6.8% solution (at zero pressure; the solubility goes up to about 9.5% at 10 bar). Therefore we have a Fermi liquid whose concentration can be varied over a wide range. Experiment shows that the heat capacity (and many other properties) follow the ideal gas model very well indeed, but with modified parameters as expected from the Landau methodology. For example, the effective mass of a ^{3}He quasiparticle in a very dilute ^{3}He–^{4}He solution is about 2.3 times the bare ^{3}He mass. This enhancement comes dominantly from interaction with the ^{4}He background for solutions with less than 1% concentration, so that it is independent of the ^{3}He concentration. These solutions are a beautiful example of the detail of FD statistics, since the transition from classical (MB) towards quantum (FD) behaviour can be followed from well above to well below the Fermi temperature. The Fermi temperature of a 0.1% solution is about 27 mK, and varies as (concentration)$^{2/3}$ (see section 8.1.3), so there is a lot of measurable ^{3}He physics between 100 mK and (say) 5 mK, achievable with a dilution refrigerator (which itself operates on ^{3}He–^{4}He solubility, as mentioned earlier in section 10.1.1).

14.3 LIQUID HELIUM-4: A BOSE LIQUID?

Unlike the previous two sections about Fermi liquids, this section will be short. This is because there is no simple way of dealing with interactions in a Bose–Einstein system. In the Fermi–Dirac case, the effect of the 'unfriendly' statistics is to keep the particles apart. This limits the scope of the interactions and allows Landau theory to be used. There is some reasonable separation between one particle and the 'others' which dress it to form a quasiparticle, simply because those 'others' are all in different states. Hence the success of the quasiparticle idea. In the case of a degenerate BE gas, there is no such separation. In fact, the 'friendly' statistics encourage precisely the opposite, and the particles in the Bose-Einstein condensation crowd into the ground state.

What we have seen in the experimental situation (section 9.2) is that the liquid does have a phase transition, but one of a markedly different behaviour from that of the non-interacting ideal gas. This contrasts with the Fermi liquids, in which the simple gas theory still works accurately, so long as a few parameters are 'adjusted'. For the Bose liquid, the whole temperature dependence is modified. The transition to superfluidity in ^{4}He is a λ-transition, rather than the much gentler third-order transition (compare section 11.1) at the 'condensation temperature' of an ideal BE gas. It seems that the interactions enhance the co-operative behaviour of the system near the transition. (This is reminiscent of our remarks about the transition to ferromagnetism in real materials which have short-range interactions (section 11.3) as opposed to the ideal mean field (long-range) interaction of section 11.2. Co-operation between noninteracting bosons by statistics only is an extreme example of a long-range effect.)

There is another mystery about liquid ^{4}He, thought to arise from the interactions. In the Bose–Einstein gas and in real liquid ^{4}He one can explain the low temperature behaviour in terms of a two-fluid model (see section 9.2). As T approaches zero, the whole material is seen to behave as a pure superfluid, i.e. there is no normal fluid remaining. As described in the previous section, the ^{4}He is thermally dead. In the theoretical friendly BE gas, the superfluid is pictured simply as having all particles together in the one-particle ground state. That is the Bose–Einstein condensation. Thus it is natural and reasonable to visualize the pure superfluid ^{4}He in the same way, as pure 'condensate'. But actually, it seems that interactions have a much more subtle effect. Two experiments using entirely different techniques (one involving neutron scattering and the other surface tension) have measured the 'condensate fraction' in the pure superfluid, and agree that it is about 13–14% only of the total particle number. It seems that the interactions actually deplete the one-particle ground state and that most of the ^{4}He atoms are scattered into other states, but that the occupation of these other states is not observed in any normal transport or thermal experiments. A convincing physical explanation of this strange state of affairs remains as yet undiscovered, it seems.

14.4 REAL IMPERFECT GASES

In the last section of this chapter, we return to ordinary honest gases, like the air we breathe. So far we have made two simplifications to the treatment of real gases. One is that, for all but low-temperature helium gas, the dilute MB limit applies, i.e. the quantum nature of the gas has no practical effect. A classical treatment applies. This is a very good approximation, and one that we shall continue to use and to exploit. The second simplification has been to ignore interactions. This is all right for dilute enough gases, where the molecules are distant, and it is well known that ideal gas theory is accurate and universal in this limit. However, it is equally well known that gases liquefy at high enough densities and low enough temperatures, and that here interactions play the dominant role. The idea of studying the statistics of imperfect gases is to see the onset of interaction effects, starting from the ideal gas side.

We shall do this in two stages. Step one is to formulate statistical physics in a classical manner – this has great interest, in that the whole subject predated quantum theory, so that we are getting back to Boltzmann. Step two is then to start to apply the classical treatment to an imperfect gas, although we shall give up when the going gets too technical.

14.4.1 Classical statistical mechanics

The classical approach to statistical physics was to use the concept of *phase space*. Consider first the state of one particle, a gas molecule in a box. Its classical state is specified by the position and momentum of the particle, something which Heisenberg's Uncertainty Principle nowadays tells us can only be done with limited precision. Phase space is defined as a combined position and momentum space. For the phase space of one particle, there are six dimensions to the space, and the state of the particle is then specified by a single point in the space. For a quantum mechanic, there would be worries if the point were not cloudy (uncertain in position in phase space) over a volume of about h^3. This is because we expect $\Delta p_x \, \Delta x \sim h$ and similarly for the y and z pairs of dimensions, giving a six-dimensional volume uncertainty of h^3.

It is instructive to see how this idea ties in with our discussion of the quantum states for a particle in an ideal gas in Chapter 4. There we saw that the states could be represented by points in k-space with a separation of $2\pi/a$ (a is a cubical box dimension and periodic boundary conditions are used). This gave an equal volume occupied in k-space for every state of $(2\pi)^3/V$, where $V = a^3$ is the volume of the cube. Hence the result of Chapter 4: there are $V/(2\pi)^3$ states per unit volume in k-space, whatever the size and shape of the real space volume V.

How does this translate into phase space? Consider a small volume in phase space $\delta v = \delta p_x \, \delta p_y \, \delta p_z \, \delta x \, \delta y \, \delta z$. This is the product of a small volume in real space ($\delta x \, \delta y \, \delta z$) and an element of momentum space ($\delta p_x \, \delta p_y \, \delta p_z$). But, using de Broglie's relation $p = \hbar k$, the momentum space volume is simply $\hbar^3$ times the k-space volume. Therefore, using the k-space result above, we can at once work out that the six-dimensional volume δv contains a number of states equal to $\delta v/[\hbar^3 (2\pi)^3] = \delta v/h^3$.

Thus, as suggested from hand-waving earlier, each and every h^3 volume of phase space has one quantum state associated with it. This is a remarkably simple geometrical result.

The classical treatment for the statistical physics of an ideal gas then proceeds as follows. The microstate (state of the assembly of N particles) can be specified by a single point in a $6N$-dimensional phase space; this point defines the momentum and position of each and every particle. The energy constraint of the microcanonical ensemble means that the representative point must lie on a $(6N-1)$-dimensional constant energy 'surface' in this space. The old *averaging postulate* was to give *equal weight to equal 'areas'* of this surface. From a suitable generalization of discussion in the previous paragraph, it is now easy to see that this use of volumes in phase space as the statistical weight for classical microstates is identical to the quantum assignment of equal weight to each individual microstate. The connection is that equal volumes h^{3N} of phase space each contain one quantum state.

The end result is that in the classical limit we simply replace sums over states by integrals over the appropriate phase space. It is worth again stressing that in the prequantum era of Boltzmann, it was a monumental achievement to make the above averaging postulate.

As one example of the classical approach, consider again the calculation of the one-particle partition function Z for an ideal monatomic gas. In Chapter 6 we used ideas of k-space to do the sum over states as an integral in k (see (6.4)) to give finally the result (see (6.6)) $Z = V(2\pi Mk_BT)^{3/2}/h^3$. The classical partition function is obtained as a simple integral of the Boltzmann factor $(\exp -\varepsilon/k_BT)$ over the one-particle phase space

$$Z_{\text{class}} = \int \exp[-p^2/(2Mk_BT)]\,dp_x\,dp_y\,dp_z\,dx\,dy\,dz$$

Since the energy does not depend on position, the space part of the integral simply gives a factor V and the momentum part is treated the same way as was the k integral in Chapter 6. The result is that $Z_{\text{class}} = V(2\pi Mk_BT)^{3/2}$, i.e. the same as before, only without the constant factor h^3, which we now know is needed to convert phase space to quantum states.

14.4.2 Imperfect gases

We are now ready to look at imperfect gases. One can realize immediately that the difference to be taken into account is that (unlike in the previous paragraph) the energy is now dependent on molecular position as well as momentum. Therefore we must work in the conceptual framework of the canonical ensemble, which involves the assembly partition function Z_A (see (12.5)), worked out from all possible energies of the whole assembly of N gas particles. As explained in section 12.2, if we first work out Z_A, then we can compactly and conveniently compute thermodynamic quantities. Here we are particularly concerned with the pressure, in order to quantify the corrections to the equation of state for an imperfect gas when interactions are switched on. What we shall expect to obtain is a result of the form

$$\frac{P}{k_BT} = n + B_2n^2 + B_3n^3 \cdots \tag{14.4}$$

where $n = N/V$ is the number density of the gas. Equation (14.4) is called the *virial expansion* of the equation of state, and the B coefficients are functions of temperature only. The first term alone gives $PV = Nk_BT$, the ideal gas equation of state.

The attempt to calculate Z_A proceeds as follows. Suppose that there is a non-zero intermolecular potential energy $\Phi(\boldsymbol{r}_1, \boldsymbol{r}_2 \ldots \boldsymbol{r}_N)$ between the N molecules. For a gas, one can safely assume that this potential energy is derived from a sum over all pairs of molecules of the pair potentials, $\phi(r_{ij})$ being the potential energy of a pair of molecules a distance $\boldsymbol{r}_{ij} = \boldsymbol{r}_i - \boldsymbol{r}_j$ apart.

The partition function Z_A is defined as the sum over all possible assembly states of the quantities $\exp(-E/k_BT)$ where E is the energy of the assembly state. Thus it is evaluated as the integral

$$Z_A = \frac{1}{h^{3N}}\frac{1}{N!}\int \exp(-E/k_BT)\,d\boldsymbol{p}_1\ldots d\boldsymbol{p}_N\,d\boldsymbol{r}_1\ldots d\boldsymbol{r}_N \tag{14.5}$$

We use the contracted notation $d\boldsymbol{r}_i$ for $dx_i\,dy_i\,dz_i$, the volume element for the ith particle, and similarly for its momentum element $d\boldsymbol{p}_i$. The factor h^{3N} is the phase space to states factor mentioned above. The $N!$ factor comes from the indistinguishability of the identical gas molecules, since without it the same state would be counted $N!$ times in the integral.

The assembly energy is worked out as the sum of kinetic and potential energy, i.e.

$$E = \sum_i \frac{p_i^2}{2M} + \Phi(\boldsymbol{r}_1, \boldsymbol{r}_2 \ldots \boldsymbol{r}_N) \tag{14.6}$$

When (14.6) is substituted back into (14.5), the integral splits into two parts, a space part and a momentum part. The momentum part is precisely the same as that for an ideal gas, and gives the product of $3N$ identical one-dimensional integrals of the form

$$\int_{-\infty}^{\infty} \exp(-p^2/2Mk_BT)dp$$

where p is a momentum component of one particle. This integral is readily evaluated (I_0 in Appendix C) as $(2\pi Mk_BT)^{1/2}$. Hence the partition function, (14.5), can be written as

$$Z_A = \frac{1}{h^{3N}}\frac{1}{N!}\cdot(2\pi Mk_BT)^{3N/2}\cdot Q \tag{14.7}$$

where Q is called the *configuration integral*, defined as

$$Q = \int \exp(-\Phi/k_BT)\,d\boldsymbol{r}_1, \boldsymbol{r}_2\ldots d\boldsymbol{r}_N \tag{14.8}$$

It is an integral over the whole of a $3N$-dimensional real space giving the positions of all the N particles (so-called configuration space). We may note at once that for a perfect gas, i.e. one for which Φ is zero for all particle positions, the integrand in (14.8) is unity and hence $Q = V^N$. We thus recover the usual result (see section 12.2): $Z_A = Z^N/N!$, with $Z = V(2\pi Mk_BT/h^2)^{3/2}$.

In the interacting situation, the configurational integral depends on the relative positions of the particles. The way the story now unfolds is to concentrate on an interaction function f_{ij} between a pair of particles i and j, defined by $f_{ij} = [\exp(-\phi(r_{ij})/k_BT)-1]$. This function is arranged so that it equals zero at large separations, it is positive at moderate separations (when the molecules attract) and it equals −1 when the molecules are very close (when the particles' hard cores exclude each other). Equation (14.8) then becomes

$$Q = \int \prod_{\text{all pairs}} (1+f_{ij})d\boldsymbol{r}_1\ldots d\boldsymbol{r}_N \tag{14.9}$$

The extended product is not so easy to evaluate after the leading term, which gives simply unity and hence a contribution to Q of V^N as for the perfect gas considered above. The interactions are all in the f_{ij}s. The terms involving f factors are often grouped using the idea of 'clusters', which specify the number of interacting particles which are involved together. The order of a cluster is chosen to be equal to the number of interacting particles minus 1. For instance, in the expansion of (14.9) there are terms with just one f factor (f from one bracket, with factors of 1 from every other bracket, e.g. f_{12}). These are single first-order cluster terms, since only one pair of interacting particles is involved. We can (given the form of ϕ_{ij}) in principle evaluate such terms. Since $f_{ij} = 0$ outside a typical atomic diameter, it is not difficult to see that each such term contributes to Q an addition of order $V^{N-1}b$ where b is the molecular volume. Next consider terms with two f factors. There are two types of these, for example $f_{12}f_{45}$ and $f_{12}f_{23}$. The first of these is a product of two separate first-order clusters, whereas the second corresponds to a single second-order cluster. These terms similarly each generate additions to Q of order $V^{N-2}b^2$.

Thus one can see that a power series of the form $Q = V^N[1+a_1(b/V)+a_2(b/V)^2+\cdots]$ emerges. This expansion is one in which b is independent of V, T and N, and in which we can expect the coefficient a_1 to be proportional to N (the number of single f terms in the expanded product) and to depend on T through the exponent $\exp(-\phi/k_B T)$. Therefore when we come to evaluate the pressure, using $P/k_B T = (\partial \ln Z_A/\partial V)_{T,N}$ (compare (12.7) and remember $P = -(\partial F/\partial V)_{T,N}$), we recover an expansion of the form of the virial expansion, (14.4).

Actually, the whole evaluation of the configurational integral gets very tough and technical, and we shall not pursue it further here. However stretching on the imagination is the detail, this cluster classification nevertheless does enable calculations to be made which give useful results for imperfect gases and which give reliable values from first principles for the virial B coefficients of (14.4).

15

Statistics under extreme conditions

In most of this book, we have been dealing with conventional materials. However, statistical physics can be useful in a number of situations where the environment is anything other than ordinary.

One of these situations concerns superfluidity. Superfluid states are normally associated with extremes of low temperature, although nowadays the advent of 'high' temperature superconductivity brings them nearer to our direct experience. We have already covered quite a lot about BE superfluids and superfluidity in liquid ^{4}He in earlier chapters (9 and 14). But more surprising, and more exotic, is the existence of superfluid states in FD systems, and this is the topic of section 15.1. At the other end of the spectrum, there are surprisingly important roles for statistical physics in our understanding of the stability of white dwarf and neutron stars, and these are discussed in section 15.2, together with a few comments about cosmology and the big bang.

15.1 SUPERFLUID STATES IN FERMI–DIRAC SYSTEMS

In our treatment of gases, we have seen that an ideal BE gas (Chapter 9) at low temperatures becomes ordered by having all of its particles enter the same ground state. The assembly ground state (at $T = 0$) is simply the symmetric combination which has all N individual particles in the same, lowest one-particle state. As we have seen, this type of coherent occupation of a single quantum state by all the particles gives a valuable (if imperfect, Chapter 14) picture of the superfluidity observed in liquid ^{4}He below 2.17 K.

An FD gas becomes ordered in a very different way, with the assembly ground state being dominated by the Pauli exclusion principle. Since all one-particle states can now be at most singly occupied, the assembly ground state (Chapter 8) has full occupation for the one-particle states with energies up to the Fermi energy $\mu(0)$, and zero occupation for those with higher energies. This provides a well-ordered (zero entropy) state, as is appropriate to $T = 0$, but the state is one of considerable energy, equal to $\frac{3}{5}N\mu(0)$ for an ideal gas. Not altogether surprisingly, nature seems to seek a way to lower this high zero-point energy. One such way is for a magnetic transition to occur, as happens in some metals. But another way is for a transition to a superfluid state to take place, and this is the topic of the present section.

For a full discussion of superfluidity in FD systems, the reader must consult a specialist book on low-temperature physics (such as the author's *Basic Superfluids*). The following will only touch upon a few aspects of what is an exciting field for both physics and technology.

15.1.1 Superconductivity

Many but not all metals become superconducting at low enough temperatures. The transition temperature T_C is typically a few kelvin (for example 1 K for aluminium, 3.7 K for tin, 7.2 K for lead; none of the monovalent metals become superconducting). Below T_C, a two-fluid model can again be used (as in superfluid ^{4}He) to describe the properties. The super-state has zero entropy, i.e. it is fully ordered. The normal fluid density varies with temperature as $\exp(-\Delta/k_B T)$, where Δ is a characteristic energy, called the energy gap, of order of magnitude $1.75\,k_B T_C$. This exponential Boltzmann factor means that the normal fluid is totally insignificant at temperatures below 0.1 or 0.2 T_C. So we can understand the temperature dependence of the specific heat and of the entropy, as illustrated in Fig. 15.1.

How does this come about? Well, a glance at the energy scale should warn that it is a subtle effect. In a typical metal, the Fermi temperature of the conduction electrons (Chapter 8) is 50 000 K, and the lattice vibrations have a characteristic temperature of order 300 K, whereas the transition temperature is merely 1 K. The idea is that there is a small *attractive* interaction between conduction electrons, arising using the intermediary of the lattice phonons, which can win out over the obvious Coulomb repulsion. This is not so unlikely when we recall the discussion of section 14.1, where we saw that at long range this repulsion is screened out by the ions. In a superconductor, electrons team up in pairs (called Cooper pairs after the originator of the idea) moving in opposite directions and with opposite spins. In jargon, these are $L = 0$, $S = 0$ pairs, where L gives the total angular momentum and S the total spin. (If you are an expert on quantum mechanics of identical particles, you can check that this arrangement will give an antisymmetric wave function as required to describe fermions!) To be oversimplistic about it, it is clear that two odds must make an even; two spin-$\frac{1}{2}$ fermions paired must give a boson, so that if the electrons conspire to occupy states only in pairs, they can fool the symmetry of the system.

The nature of the interaction between electrons is a second-order one, in that a third particle is involved. Roughly, what happens is that one electron (wavevector $\boldsymbol{k}$) in travelling through the lattice can excite a short-lived ('virtual') phonon; this lattice disturbance is then sensed in a coherent way by the second electron (wavevector $-\boldsymbol{k}$) which is travelling in exactly the opposite direction (to give $L = 0$). It is this interaction by which the conspiracy to paired occupation was found (by Bardeen,

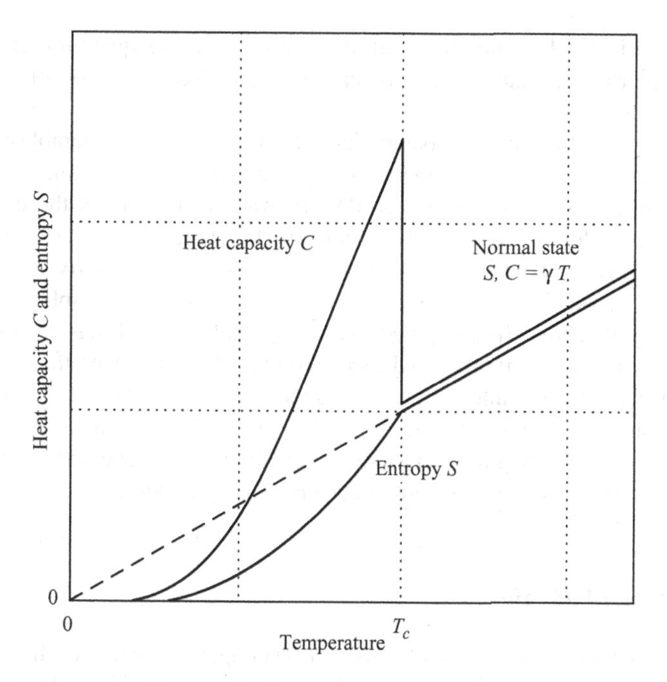

Fig. 15.1 The normal-to-superconducting phase transition in zero applied magnetic field. The graphs show the dependence on temperature of the electronic heat capacity C and entropy S. Note the increased order (decreased S) of the superconductor compared to the normal state at the same temperature. The curves are related by $C = T(\mathrm{d}S/\mathrm{d}T)$.

Cooper and Schrieffer (BCS) in 1957) to lower the energy of a coherent ground state. The energy gap Δ is the minimum energy per electron needed to break one of these pairs. The $\exp(-\Delta/k_\mathrm{B}T)$ form of the thermal properties at low temperatures follows from the basic statistical formulae (actually from the high energy tail of the FD distribution function) when it is appreciated that there is this energy gap between the ground state (superfluid) and the excited states (normal fluid).

The electrons in the coherent ground state have fascinating, somewhat alarming, properties. They are described by a single wave function (analogous to the boson ground state which describes all particles in the BE gas). This results in three classes of remarkable property: (i) the electrical resistance is zero – compare the viscosity in ^{4}He; (ii) magnetic B-fields are excluded from the metal, shielded out by surface persistent currents; (iii) there are quantized states, corresponding to magnetic flux linked in units of $h/2e$, around a loop in the superconductor. In view of the second property, it is not surprising that the superconducting state is suppressed by high enough applied magnetic fields, and the use of superconductors to produce high-field, stable and loss-free magnets is therefore a difficult problem and involves special materials. The third property gives the basis for the whole ideas of 'SQUIDs' (superconducting quantum interference devices), the most sensitive electro-magnetic measuring instrument in existence. It is worth mentioning that the 2 in the $h/2e$ flux quantum arises directly from the pairing mechanism, and is an experimental confirmation of the unlikely ideas of BCS!

In recent years, the topic of superconductivity has aroused considerable new interest with the discovery of 'high-temperature superconductivity' in certain oxide materials. Transition temperatures around 100 K are available in some of these remarkable materials, and a lot of study has been made of YBCO (yttrium barium copper oxide) and similar layered materials, in which it seems that superconductivity is associated with copper oxide planes in the structure. These oxides are miserable conductors in the normal state above T_C compared to conventional metals. It has been shown that Cooper pairs are again involved in the superconductivity. However, the precise mechanism for the attractive interaction is a matter of much controversy and uncertainty (a joke about N theoretical physicists producing $N(N-1)/2$ theories is perhaps in order here?). It is exciting to meet a technologically important field in which the basic physics is insecure, so there is a lot of current activity in this area.

15.1.2 Superfluid ^{3}He

The other Fermi–Dirac system of interest is (yet again) liquid ^{3}He. In Fig. 8.5, we introduced the ^{3}He phase diagram, which illustrates that at reasonably low pressures there is no solid phase. Instead liquid exists down to the lowest temperatures, and it is of relevance to be curious about the nature of the liquid ordered state as the absolute zero is approached. In fact it was discovered in the early 1970s that, right down in the millikelvin range, the liquid does become a superfluid. The low-temperature phase diagram is shown in Fig. 15.2.

Again, the reader must be referred to specialist low-temperature texts for detail, but a brief outline of some salient points follows.

1. Cooper pairs (two odds make an even!) are again involved. The mechanism is as before a second-order process, but the intermediary is now not the phonon system. Rather it is the magnetic polarization of the background ^{3}He fluid. As we noted in section 14.2.2 the magnetic moment of one spin-$\frac{1}{2}$ ^{3}He atom interacts strongly with the others. Hence in the second-order process, one half of a Cooper pair leaves a polarization trail in the surrounding fluid and this is then sensed by the second half of the pair. This effect is a subtle one, which is why T_C is in the millikelvin range rather than being of order kelvin (as it is in ^{4}He).
2. However, it is much more exotic than most (if not all) superconductivity. The pairs are between parallel spins, not antiparallel ones. Thus $S = 1$ and $L = 1$ in superfluid ^{3}He. This gives an extra complexity and a unique interest to this topic. The complexity is witnessed by the existence of two very different phases (the A- and the B-phases ...guess which was discovered first!) as shown in Fig. 15.2. The ordering in the ^{3}He superfluid needs to be described by vector quantities, rather than the simple scalar Δ of the superconductor.

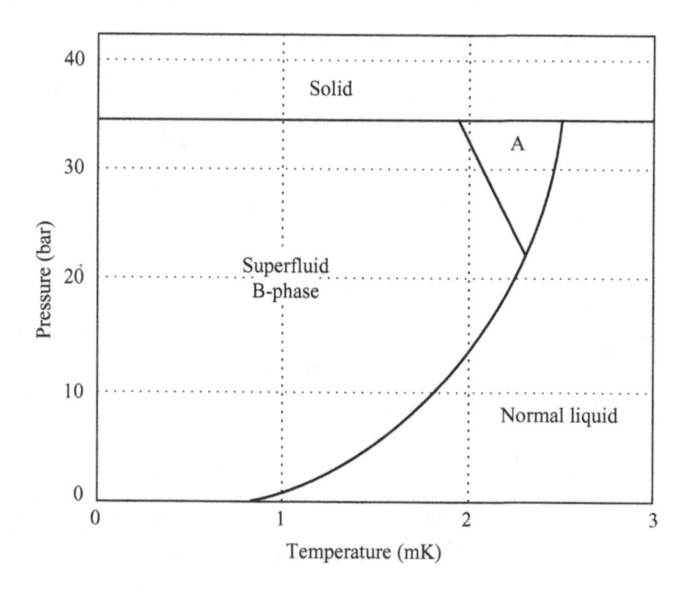

Fig. 15.2 The phase diagram of ^{3}He at millikelvin temperatures, and in zero applied magnetic field. The A- and B-phases are different superfluid phases.

3. In the A-phase, these vectors correspond to a spin (S) direction and an orbital (L) direction, shared by all pairs in a particular region of the fluid. Hence the A-phase has many similarities with liquid crystals, which also order with a highly anisotropic vector ordering parameter. (The magnitude of the order parameter gives the strength of the ordering; the direction of the vector gives the direction of alignment of the liquid crystal molecules.)
4. In the B-phase, the $S = 1$, $L = 1$ pairs collect together in a much more uniform way, and the anisotropy is practically non-existent (although it can have secondary manifestations). Many experimental properties of the B-phase are described simply in terms of a single scalar Δ parameter. Figure 15.3 gives an example of this type of behaviour, and it also serves as a dramatic illustration of the existence of superfluidity. As noted above in section 14.2.2, ^{3}He is a highly viscous liquid above T_C, so that the damping of a vibrating wire is very large. When the liquid is cooled below T_C, however, the damping falls off rapidly, varying as $\exp(-\Delta/k_B T)$, the usual Boltzmann factor, at low temperatures as the normal fluid is frozen out. At the lowest temperatures the vibrator behaves as if in a vacuum with a damping coefficient 100 000 times less than at T_C.
5. The fact that the pairs have $S = 1$ means that they are magnetic. The behaviour of the superfluid is thus profoundly influenced by applied magnetic fields. This is so even in the B-phase, which becomes unstable if a large enough field (of order 0.4 T) is applied, the superfluid reverting to the more magnetic A-phase. The whole phase diagram changes, which is why the $B = 0$ qualification was needed

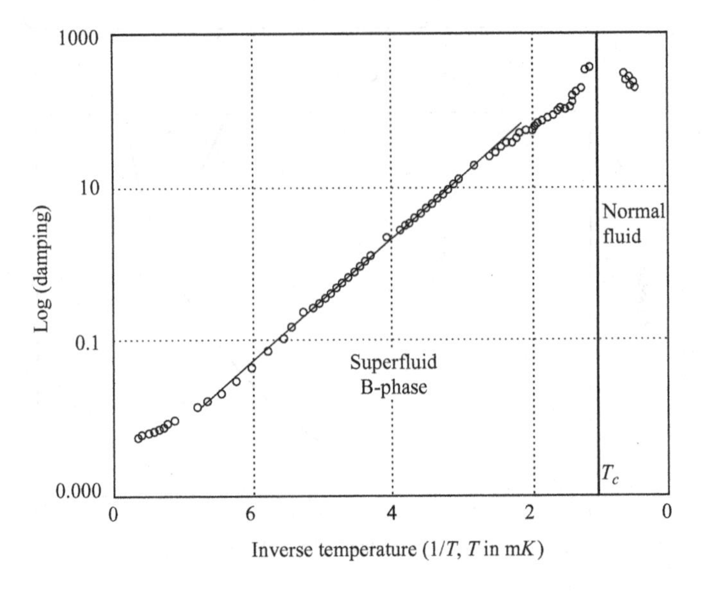

Fig. 15.3 The damping of a vibrating wire in liquid ^{3}He, measured by the author and colleagues in Lancaster. The damping changes by about five orders of magnitude from the transition down to the lowest temperature (130 μK in this work). The damping is plotted logarithmically against $1/T$, and the linear relation in the superfluid shows that the damping is frozen out by the gap Boltzmann factor, $\exp(-\Delta/k_B T)$.

in Fig. 15.2. In contrast with superconductors, the effect of the field on T_C itself is minor, but it reinforces one phase rather than the other. In the usual low-field A-phase the B-field has a marked effect on the vector ordering by a tendency to align the local S direction. Since in the A-phase L is itself constrained to be perpendicular to a wall (otherwise the pair would rotate into it), and there is also a dipole coupling between spin and orbital motion, the whole system is one in which topological insight and imagination are required. Thus superfluid ^{3}He gives an accessible laboratory for all sorts of theoretical speculation in a whole variety of different regions of physics, from cosmology to turbulence.

15.2 STATISTICS IN ASTROPHYSICAL SYSTEMS

The simple ideas of statistical physics turn out to have a profound influence on our understanding of some parts of astrophysics. We shall briefly discuss two such areas (out of many). In the first, according to modern (fashionable?) theory, a simple Boltzmann factor turns out to be a vital factor to tell us about the chemical makeup of our universe shortly after the big bang. The second area is at the opposite end of the time-scale, in our understanding of the stability or otherwise of certain types of stellar matter. The properties of dense Fermi-Dirac gases are a basic part of a discussion of white-dwarf stars and of neutron stars, and we shall outline some of the features.

Whether or not superfluidity is a part of the understanding involves speculation too uncertain even for the average astrophysicist, and we will not be drawn into this intriguing question!

15.2.1 After the big bang

The nowadays orthodox view of big bang cosmology is based on three principal pieces of evidence: (1) the observation by US astronomer Edwin Hubble and others that the universe is expanding, (2) the existence of the cosmic background radiation, and (3) the observation that 25% of the mass of the universe is helium.

The idea is that, since the universe is now expanding, we can imagine running time backwards towards a beginning, where the universe was effectively at a point and had infinite temperature. Forward from that (almost unimaginable) big bang, we have a system which is gradually expanding and cooling, with the 'normal' laws of physics applying after about time $t = 10^{-34}$ seconds. At that stage, the temperature was about $T = 10^{27}$ K.

At the present day, $t = 15 \times 10^9$ years, the temperature has cooled to that of the uniform cosmic background radiation. Recent measurements from the COBE satellite have confirmed that our universe today is filled with rather (but not exactly) spatially uniform radiation. The spectral density of the radiation follows very precisely the black-body radiation spectrum (see (9.13) and Fig. 9.7) corresponding to a thermal equilibrium temperature of 2.726 K. This is seen as a logical legacy of the original hot big bang and was predicted long before its eventual measurement. It is a very beautiful example of the photon gas treatment of Chapter 9.

The origins of the 25% helium involves many of the ideas introduced in our discussion of chemical reactions in Chapter 13, and in particular concerns Boltzmann factors of the type $\exp(-\Delta E/k_{\mathrm{B}}T)$. Consider first what happens up to about $t = 10^{-3}$ seconds, by which time the temperature T has cooled to a mere 10^{12} K. The universe will contain a lot of photons (γ rays) of typical energy several times $k_{\mathrm{B}}T$ (about 100 MeV), using the black-body spectrum. This means that there will also be created particles and antiparticles of various sorts (thought to be quarks and other exotic particles, the 'quark soup'). As the system cools, so does the energy scale of the photons and hence also the upper mass of creatable particles (using $E = mc^2 \sim k_{\mathrm{B}}T$).

By $t = 10^{-2}$ seconds, the temperature has fallen to around 10^{11} K, and the energy scale to 10 MeV. This is now too cold for anything but the particles with which we are familiar: photons, electrons, neutrinos, neutrons and protons. Nucleosynthesis (manufacture of heavier nucleons) has not yet started, since ample γ photons are available to break up any transiently formed deuterons. Thermal equilibrium prevails, hence the interest to statistical physics. Because of this equilibrium, the neutron/proton ratio is determined simply by the temperature. We would predict (correctly) that, in thermal equilibrium, the number of neutrons divided by the number of protons is

$$N_n/N_p = \exp(-\Delta mc^2/k_{\mathrm{B}}T) \tag{15.1}$$

where Δm is the mass difference between neutron and proton. Using the known mass values, we find that Δmc^2 is about 1.3 MeV (the proton mass is 938 MeV). Thus $\Delta mc^2 = k_B T$ at about 1.5×10^{10} K. Thus (15.1) suggests that at 10^{11} K, the numbers of neutrons and protons are substantially equal; we are in a high-temperature limit for this equilibrium. At 10^{10} K, the ratio should be about 0.22, and at later times (lower temperatures) it should rapidly drop to zero according to (15.1). However, here we can use our knowledge (Chapter 13) of reactions to realize that the thermal equilibrium situation will not last for ever. There is an activation energy for the beta-decay reactions which maintain equilibrium between neutrons and protons. What happens is that at around 10^{10} K ($t \sim 1$ second), the reactions become sufficiently slow that the approximately 20% of neutrons are effectively frozen out in a metastable state on the current time scale.

In the next period, nucleosynthesis is still delayed by the high energy tail of the gamma photons. However by about 100 seconds (10^9 K) these photons are sufficiently few to allow deuterons and hence other nuclei to start to form. Many of the neutrons thus end up as helium, the alpha particle being a very stable and thus favoured product. Modelling suggests a few other nuclei, including a very little ^{7}Li. By 10 000 seconds (10^7 K, 10 keV) there are no free neutrons left, since those few which did not form nuclei have in any case beta-decayed spontaneously into protons (half-life = 1000 seconds). Thus no further nuclei are formed, since two nucleons must overcome a Coulomb barrier to fuse, and the thermal energy is not large enough to allow this. Thus no heavy nuclei are formed at this stage, and their formation must await the nuclear reactions in the hot centres of stars millions of years later. We may note that the suggestion of essentially 25% helium, 75% hydrogen for the nuclei at this stage implies a neutron/proton ratio of 1/7, entirely consistent with the above scenario.

15.2.2 The stability of stars

Astrophysicists have always been good at making up names. The majority of observed stars (of which our Sun is one) show a correlation between brightness and colour. These form the 'main sequence' stars, in which faint stars are red and bright stars are white. But there are exceptions, with the picturesque names of red giants (and supergiants) and white dwarfs.

In the main sequence, the energy source is primarily from hydrogen fusion reactions to produce helium, generating and sustaining high temperatures. The correlation relates to the black-body radiation spectrum from a hot body. The hotter it becomes, the brighter it is ($U \propto T^4$) and the whiter it is ($\nu_{\max} \propto T$) as shown in section 9.3.1. Thus the radiation from our Sun is essentially that of a black body with temperature equal to that of the surface Sun temperature (about 6000 K).

An interesting question is what happens to a star when the hydrogen is finally burnt out. How does it die? Certainly initial cooling and gravitational collapse can then occur. But within this context there are (at least) three possible next stages. One is formation of a white dwarf, one is formation of a neutron star and the third is collapse in a single stage to a black hole. Both of the first two of these involve highly dense matter and extremely degenerate Fermi–Dirac gases.

First, consider white dwarf stars. They are faint objects, because the principal energy source is simply from a gradual gravitational collapse. Nevertheless, by becoming dense they remain hot, with a typical core temperature of 10^7 K, similar to that of the core of the Sun. Hence the white colour. They are made up of helium (and/or of heavier nuclei – this is not known, and it has little effect on the following). They are extremely dense, about 10^7 to 10^{11} kg m^{-3}, i.e. a million times more dense than the Earth (imagine your body weight in a cubic millimetre!). At these temperatures, the electrons are all free from the nuclei, so that there is a dense FD gas of electrons. Coulomb forces ensure that the material remains overall neutral, the electrons follow the nuclei and vice versa. The stability of the white dwarf arises from a balance between the attractive gravitational potential energy of the nuclei and the high kinetic energy (and hence pressure) of the FD electron gas. It turns out that a heavy white dwarf is never stable, the largest possible mass being about 1.4 solar masses for it not to suffer further gravitational collapse.

In a neutron star, further collapse of the material leaves our neutral star all in the form of neutrons, the protons and electrons having combined by inverse beta-decay. The density is even higher than for a white dwarf star, around 10^{13}–10^{17} kg m^{-3}. This means that the neutrons are only a few neutron radii apart, not far from pure nuclear matter. Again stability of the star occurs when there is a balance between gravitational attraction and the outward pressure of the highly degenerate FD neutron gas. Once again it turns out that there is an upper mass limit for stability, namely 1.5–2 solar masses although an exact calculation is not easy. A larger star must, it seems, collapse to become a black hole in a single step.

Let us briefly examine the stability of these systems in more detail.

Pressure of a degenerate FD gas. In section 8.1, we worked out the pressure of a uniform, non-relativistic, spin-$\frac{1}{2}$, ideal FD gas. There are three steps:

(a) The Fermi wavevector is readily shown (see (8.6)) to be equal to $k_F = (3\pi^2 N/V)^{1/3}$. This is just a bit of 'waves-into-boxes' geometry.
(b) The internal energy of the gas is worked out using the (non-relativistic) dispersion relation $\varepsilon = \hbar^2 k^2/2m$. (Note: we use the symbol m for the particle mass in this section, in order to reserve M for the mass of a star). The answer at $T = 0$ (see (8.8)) is $U = \frac{3}{5} N\mu$, where μ is the Fermi energy (= $\hbar^2 k_F^2/2m$).
(c) Finally, the pressure is worked out (section 8.1.3) from $P = \frac{2}{3} U/V$, where the $\frac{2}{3}$ factor also arises from the dispersion relation (since $\varepsilon \propto V^{-2/3}$). The pressure is thus given by

$$P = \frac{2}{5}\frac{\hbar^2}{2m}(3\pi^2)^{2/3}(N/V)^{5/3} \qquad (15.2)$$

In the case of a white dwarf star, the electron gas can be relativistic. It is instructive to follow the same steps as before, and to realize that the calculation generalizes to include relativity with rather few problems. Step (a) is identical to the above, since it involves k-space geometry only. The expression for k_F is thus unchanged. However, in steps (b) and (c), we must allow for appropriate modifications to the dispersion relation. For a particle of rest mass m and momentum $\hbar k$, the energy–momentum relationship is $\varepsilon^2 = (mc^2)^2 + (c\hbar k)^2$. In the extreme relativistic limit, this simplifies to $\varepsilon = cp$, the same dispersion relation as for photons. Hence, in the extreme limit, step (b) gives $\mu = c\hbar k_F$, leading to $U = \frac{3}{4}N\mu$. Step (c) becomes $P = \frac{1}{3}U/V$, since now $\varepsilon \propto V^{-1/3}$. Putting these results together to calculate the pressure in the extreme relativistic limit, we obtain

$$P = \frac{1}{4}c\hbar(3\pi^2)^{1/3}(N/V)^{4/3} \tag{15.3}$$

When the gas is in an intermediate regime, the full dispersion relation must be used, and some unpleasant integrals must be computed. What happens is that (15.2) and (15.3) join up smoothly.

The stability requirement. This is difficult to work out exactly, but quite easy to get the rough idea. So let us just concentrate on the rough idea! Consider first a star in the absence of gravitation. We treat the star as a number N of spin-$\frac{1}{2}$ gas particles, confined within a 'box' of volume $V = \frac{4}{3}\pi R^3$ where R is the radius of the star. In the absence of gravity, the density of the gas will be uniform. If the star were now to be collapsed by a further amount δR, an amount of external work would need to be done equal to $P4\pi R^2\delta R$, i.e. to the pressure times the volume change.

In practice, of course, this energy must be supplied by the loss of gravitational potential energy of the star. We need to turn gravity on! This is where the trouble starts for our calculation, since the density of the star is no longer uniform. If it were to remain uniform, a straightforward integration gives for the gravitational potential energy of a spherical mass the expression $-\alpha GM^2/R$, with the constant $\alpha = 3/5$. Here M is the total mass of the star. For a non-uniform star it is a rough and reasonable assumption that this expression will remain valid (dimensional analysis!), but that the true value of α might differ somewhat from 3/5. Thus the gain in potential energy for a reduction in radius of δR is equal to $\alpha GM^2\delta R/R^2$.

Our star is thus stable when there is balance between the gravitational potential energy change and the work required to combat the pressure, i.e. when

$$P = \alpha\frac{G}{4\pi}M^2/R^4 \tag{15.4}$$

Let us suppose that the non-relativistic approximation for the pressure (15.2) is valid. This turns out to be the case for neutron stars and for light enough white dwarf stars. Putting together (15.2) and (15.4) gives us the relationship between mass and size for a stable star. Writing V as $\frac{4}{3}\pi R^3$, we have

$$KN^{5/3}/R^5 = M^2/R^4 \tag{15.5}$$

where K is a known constant (roughly known that is, since it includes α). In fact, assuming $\alpha = 0.6$, simple substitution gives in SI units (i) for electrons as in a white dwarf $K = 6.6 \times 10^{-28}$, and (ii) for neutrons as in a neutron star $K = 3.6 \times 10^{-31}$.

All that remains is to relate the number N of gaseous fermions to the total mass M of the star. For a neutron star, we have simply $M = Nm_{\mathrm{n}}$ where m_{n} is the neutron mass. For a white dwarf, the relation depends in detail on the (unknown) composition; if it is helium then $M = 2Nm_{\mathrm{n}}$ effectively, since each electron has two nucleons associated with it. Any other composition gives a relation of similar magnitude. For both types of star, (15.5) thus becomes $M^{1/3}R$ = constant, i.e. mass $\times$ volume = constant. This is an intriguing result, verifying the alarming inverse relation between mass and volume. A heavy stable star of this type is *smaller* than a lighter one. The density increases dramatically as the square of the mass. This idea gives the correct order of magnitude for the properties of these stars mentioned above. Using the values of K given above, it follows that, for a stellar mass equal to that of the Sun (2×10^{30} kg), the radius of a white dwarf is about 7000 km and the radius of a neutron star is a mere 12 km. Hence the amazingly high densities.

The upper mass limit. It is not hard to believe that these results imply that very heavy white dwarfs or neutron stars cannot be stable at all. Certainly, as we have seen, the heavier the star the smaller and more dense it will become. Moreover, our treatment is only approximate, and at high enough densities other factors come into play.

In the case of a white dwarf the new factor is the high Fermi velocity of the electron gas. As relativistic speeds are approached, we need move our calculation of the pressure of the gas from (15.2) towards (15.3). It is now easy to see that a limit is involved. If we combine (15.3) with (15.4) we obtain (instead of (15.5))

$$K_1 N^{4/3}/R^4 = M^2/R^4 \tag{15.6}$$

with the constant $K_1 = 1.1 \times 10^{-15}$ in SI units.

This is another intriguing result, since the radius of the star does not appear: R^4 cancels R^4. The only moving part in (15.6) is the mass. For a white dwarf star with $M = 2Nm_{\mathrm{n}}$ as above, the equation solves for a mass $M = 3.4 \times 10^{30}$ kg, i.e. about 1.7 solar masses. The implication is that lighter white dwarfs are stable, with a radius effectively given by (15.5). As the star gets heavier, the star collapses even faster than the non-relativistic expression would indicate, until for stars heavier than 1.7 solar masses there is no stable solution; the relativistic gas simply cannot generate a high enough pressure to overcome the gravitational attraction. Hence our prediction is that the maximum size of a stable white dwarf is 1.7 solar masses. (Incidentally the correct value for this limit, when the correct density profile in a star is included is actually 1.44 solar masses, the 'Chandrasekhar limit'. Our simple treatment is not so bad!)

A neutron star also gets into problems at high enough mass and thus at extremely high density. Relativity again limits the validity of the simple treatment, but the problem now hinges on the strength of the gravitational attraction, a factor which becomes of importance well before any corrections for the relativistic Fermi velocity of the neutrons. It is all to do with the horizon threshold of a black hole. As soon as the escape velocity for a particle from the gravitational field of a star becomes equal to the speed of light, then the star must collapse to a black hole. The limiting radius for this to occur is the 'Schwarzschild radius' R_0 given by $c = (2GM/R_0)^{1/2}$, i.e. $R_0 = 2GM/c^2$. Combining this idea with (15.5) we see that the maximum mass of a neutron star is then given by

$$M^{4/3} = Km_{\mathrm{n}}^{-5/3} c^2/2G \tag{15.7}$$

Substitution gives the mass limit from (15.7) to be 2.8 solar masses. Again this is probably an over-estimate of the true value, thought to be somewhat less than 2 solar masses.

As a footnote, it is interesting to ask whether there is any firm observational evidence for neutron stars. The answer seems to lie in the discovery (by Anthony Hewish and Jocelyn Bell in the late 1960s) of pulsars. These are objects which emit radio waves (or other electromagnetic radiation) in regular bursts, i.e. pulses. The pulses have a period in the range milliseconds to seconds which is characteristic to the particular pulsar. The origin of pulsars was not immediately obvious and they were nicknamed at first 'little green men'. However, it was soon realized that highly dense stars, white dwarfs or neutron stars, must be involved. When the numbers were put into possible theories, the observational details of the emitted radiation gave convincing evidence that the likeliest candidate for a pulsar is a rapidly rotating neutron star. Such an object will have around it a remnant of electrons and an intense magnetic field (about 10^8 T, compared to our terrestrial field of 1 mT) which confines these electrons. The accelerating electrons are responsible for the emission of the radiation. Since the rotational speed of the star is relativistic, the radiation is emitted predominantly in the forward motion direction; this effect is called 'synchrotron radiation', the basis of many modern research machines designed to provide a well-collimated and intense source of radiation. The pulsar effect can now be understood as a sort of lighthouse effect as the star rotates, so long as the magnetic field axis is different from the axis of rotation.

In summary, we have seen that the application of a little statistical physics can throw much light on a number of interesting questions about stars. There remain many other questions of importance, such as whether superfluidity (Chapter 14) of the dense gas plays a role in these systems, a real possibility in neutron stars. But physics would not be an interesting subject if all the answers were easily available.

Appendix A

Some elementary counting problems

Suppose you have N distinguishable objects.

1. *In how many different ordered sequences can they be arranged?*

The answer is $N!$

The explanation is not hard, but in this case as always the reader is encouraged to test the result with a few simple examples (e.g. try arranging three objects, A, B and C, and identify the $3! = 6$ arrangements). In general you should convince yourself that in first position in the sequence there are N choices of object; in second place (having made the first-place choice) there are $N - 1$; and so on, until for the Nth place there is just one object left to choose. Hence the required result is $N \times (N-1) \times (N-2) \ldots 1$, i.e. $N!$.

2. *In how many ways can the* N *objects be split up into two piles, ordering within the piles being unimportant? The first pile is to contain* n *objects and the second* m.

The answer is obviously zero, unless $n + m = N$. If the numbers do add up, then the answer is $N!/(n! \times m!)$.

The zero answer may seem trivial, but it is analogous to the requirements of getting the total number of particles right in a statistical distribution. There are several ways of seeing the correct answer, of which one is as follows. Call the required number t. We can recognize that the $N!$ arrangements (problem 1) can be partitioned so that the first n placed objects are put into the first pile, and the remaining m placed objects into the second pile. However, these $N!$ arrangements will not all give distinct answers to the present question, since the order within each pile is unimportant.
Hence, we can see that

$$N! = t \times n! \times m! \tag{A.1}$$

where the $n!$ and $m!$ factors allow for this disordering (again using the result of problem 1). Again convince yourself with a small-number example.

Before leaving the problem, it is important to recognize that it is identical to the binomial theorem question: what is the coefficient of $x^n y^m$ in the expansion of

$(x+y)^N$? And the answer is the same: zero unless $n+m=N$, but if $n+m=N$ then the required number is $N!/(n! \times m!)$.

3. *In how many ways can the* N *objects be arranged if they are now split up into* $\mathrm{r}+1$ *piles with* $\mathrm{n_j}$ *objects in pile number* j $(\mathrm{j}=0,1,2,\ldots\mathrm{r})$*?*

The answer is zero unless $N=\sum n_j$. If the n_js do sum to N then the required number of ways is $N!/(n_0!n_1!\ldots n_r!)$.

This result follows as a straightforward extension of the proof of problem 2, equation (A.1) becoming: $N! = t \times \prod n_j!$. This important result is the one used extensively in Chapter 2.

Again one may note that this problem may be thought of (and solved) as the question of the multinomial theorem: what is the coefficient of $y_0^{n_0}y_1^{n_1}y_2^{n_2}\ldots$ in the expansion of $(y_0+y_1+y_2+\cdots)^N$? And the answer is identical.

Appendix B

Some problems with large numbers

1 STIRLING'S APPROXIMATION

Stirling's approximation gives a very useful method for dealing with factorials of large numbers. The form in which it should be known and used in statistical physics is:

$$\ln X! = X \ln X - X \tag{B.1}$$

for any large integer X. And another useful result may be obtained by differentiating (B.1) to give:

$$\begin{aligned} \mathrm{d}(\ln X!)/\mathrm{d}y &= (\mathrm{d}X/\mathrm{d}y)\{\ln X + 1 - 1\} \\ &= (\mathrm{d}X/\mathrm{d}y) \ln X \end{aligned} \tag{B.2}$$

where one assumes that X may be considered a continuous variable.

Equation (B.1) is not difficult for the interested party to prove. Consider the definite integral $I = \int_1^X \ln z \, \mathrm{d}z$. Integration by parts expresses this as $[z \ln z - z]_1^X$, which when evaluated between the limits gives:

$$I = X \ln X - X + 1$$

However, a quick inspection of a graph of this integral (Fig. B.1) shows that the integral (i.e. the area under the curve) lies almost midway between the areas of the two staircases on the diagram. And the upper staircase has an area of $(\ln 1 + \ln 2 + \cdots + \ln X) = \ln X!$.

Therefore we see that $I = \ln X!$ – an error term of approximately $\frac{1}{2} \ln X$.

The lower one's area is

$$\begin{aligned} [\ln 1 + \ln 2 + \cdots + \ln(X - 1)] &= \ln(X - 1)! \\ &= \ln x! - \ln X \end{aligned}$$

Hence: $\ln X! = X \ln X - X [+\frac{1}{2} \ln X + \text{term of order unity}]$. This is Stirling's approximation. It really is remarkably accurate for the sort of numbers we use in statistical

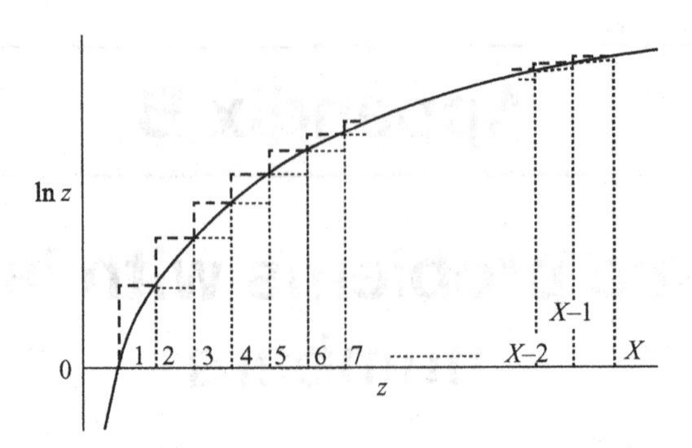

Fig. B.1 Stirling's approximation.

physics; even for a very modest $X = 1000$, the error in using equation (B.1) is less than 0.1%, and the fractional error goes roughly as $1/X$.

2 A PROBLEM WITH PENNIES

An unbiased penny is tossed N times. (i) How many possible arrangements of heads (H) and tails (T) are there? (ii) What is the most probable number of H and T after the N tosses? (iii) What is its probability? In particular what is this probability when N is large?

This problem bears much similarity to the statistical physics at high temperatures of a spin-$\frac{1}{2}$ solid (section 3.1), and we shall apply the usual nomenclature in discussing it. The answer to the first problem is simply 2^N ($= \Omega$, say, by analogy), since each toss has two possibilities. The answer to problem (ii) is also readily solved since the coin is unbiased. This means that each of the 2^N arrangements (or microstates) has equal probability; hence the most probable distribution of H and T is that which includes the most microstates. And that is the distribution with equal numbers of heads and tails (we assume N to be even for convenience!). This is readily proved from problem 2 of Appendix A. A distribution of nH and $m(= N - n)T$ can occur in $t = N!/(n!m!)$ ways, and this is maximized when $n = m = N/2$.

Problem (iii) is then also solved. The required probability P is given by t^*/Ω, i.e. $P = N!/\{(N/2)!^2 2^N\}$. However, the interest and the relevance to statistical physics come when we evaluate P when N is large. Here is the calculation:

$$\begin{aligned}
\ln P &= \ln N! - 2\ln(N/2)! - N\ln 2 && \text{from above} \\
&= N\ln N - N - 2(N/2)\ln(N/2) && \text{from (B.1)} \\
&\quad +2(N/2) - N\ln 2 \\
&= 0 && \text{since everything cancels!}
\end{aligned}$$

Hence $\ln P = 0$, or $P = 1$, within the accuracy of Stirling's approximation. Another way of stating this result is as: $t^* = \Omega$, a result we have used in Chapter 2 and elsewhere. Since we have noted the success of (B.1) we can have confidence in our statistical method.

However, before leaving this example, a word of warning. It is not in fact true that the probability of getting *exactly* $N/2\,H$ and $N/2\,T$ increases towards 1 as N increases! Rather $\ln P$ gets smaller as $\ln N$ (the error in (B.1)), as may be verified by anyone with a laptop computer and a sense of adventure. But the sharpness of the probability at the maximum does increase as N gets larger, defined say by the probability of getting a distribution within 0.1% of the most probable. In other words it is the probability of getting *nearly* the most probable which becomes overwhelmingly large, on any reasonable definition of 'nearly'. Fortunately, this is precisely the sort of situation in which we are interested in statistical physics.

Appendix C

Some useful integrals

1 MAXWELL–BOLTZMANN INTEGRALS

To calculate the properties of an MB gas, as in Chapter 6, we need to evaluate the definite integrals of the form:

$$I_n = \int_0^\infty y^n \exp(-by^2)\mathrm{d}y \tag{C.1}$$

where n is any positive integer. This can be done in three stages.

(i) Equation (C.1) may be integrated by parts to give

$$I_n = [-y^{n-1}\exp(-by^2)/2b]_0^\infty + [(n-1)/2b]I_{n-2}$$

For $n \geq 2$, the first term is zero since it vanishes at both limits, giving a simple recurrence relation between I_n and I_{n-2}:

$$I_n = [(n-1)/2b]I_{n-2} \tag{C.2}$$

For some purposes, for example the calculation of the rms speed of gas molecules, the recurrence relation contains enough information by itself. But equation (C.2) is useful in every case, since its application reduces any integral I_n to a known multiple of either I_1 or I_0.

(ii) The integral I_1 is obtained by simple integration, the result being

$$I_1 = 1/2b \tag{C.3}$$

(iii) The integral I_0 takes a little longer to evaluate. A quick method is to consider a two-dimensional problem, to integrate $\exp(-br^2)$ over the whole $x-y$ plane, r being the distance from the origin. We know from the definition (C.1) that

$$2I_0 = \int_{-\infty}^\infty \exp(-bx^2)\mathrm{d}x = \int_{-\infty}^\infty \exp(-by^2)\mathrm{d}y$$

Hence

$$\begin{aligned}4I_0^2 &= \int_{-\infty}^{\infty}\int_{-\infty}^{\infty} \exp[-b(x^2+y^2)]\mathrm{d}x\,\mathrm{d}y \\ &= \int_0^{\infty} \exp(-br^2)2\pi r\,\mathrm{d}r \\ &= 2\pi \boldsymbol{I}_1 \\ &= \pi/b\end{aligned}$$

Thus

$$I_0 = \frac{1}{2}(\pi/b)^{1/2}. \tag{C.4}$$

The three equations (C.2), (C.3) and (C.4) between them enable one to evaluate any of the required integrals.

2 FERMI–DIRAC INTEGRALS

In Chapter 8, we gave several results for the properties of an ideal FD gas in the limit $T \ll T_F$. The calculations require the (approximate) evaluation of integrals of the form:

$$I = \int_0^{\infty} [\mathrm{d}F(\varepsilon)/\mathrm{d}\varepsilon]f(\varepsilon)\mathrm{d}\varepsilon \tag{C.5}$$

The function $F(\varepsilon)$ is chosen to suit the property required, and $f(\varepsilon)$ is the FD distribution, equation (8.2). Integration by parts of equation (C.5) gives

$$I = -F(0) - \int_0^{\infty} F(\varepsilon)[\mathrm{d}f(\varepsilon)/\mathrm{d}(\varepsilon)]\mathrm{d}\varepsilon \tag{C.6}$$

where we have used $f(0) = 1$ and $f(\infty) = 0$. Usually one can choose the ('user-defined') function so that $F(0) = 0$, so we shall ignore the first term of equation (C.6). The function $(-\mathrm{d}f/\mathrm{d}\varepsilon)$ is an interesting one in the limit $k_BT \ll \mu$. It is zero except within about k_BT of the Fermi energy, and in fact behaves like a 'delta-function' with a nonzero width. (At $T = 0$ it becomes identical to the delta function.) Therefore equation (C.6) is evaluated by expanding the function $F(\varepsilon)$ as a Taylor series about μ, since only its properties close to $\varepsilon = \mu$ are relevant. The result is

$$I = F(\mu) + (\pi^2/6)(k_BT)^2F''(\mu) + \cdots \tag{C.7}$$

Note that: (i) the first term, the value of F at the Fermi level, is the only one to survive at $T = 0$; that is the delta-function property, (ii) there is no first derivative term since $(-\mathrm{d}f/\mathrm{d}\varepsilon)$ is symmetrical about the Fermi level; and hence no term linear in T, and

(iii) the first correction to the zero temperature value is proportional to T^2. If $F(\varepsilon)$ is a reasonable power law of ε, then it will have a magnitude of order $(k_B T/\mu)^2$ times the first term. The factor $(\pi^2/6)$ which enters comes from the definite integral

$$\int_0^\infty y^2 \exp(y) dy / [1 + \exp(y)]^2 = \pi^2/6$$

(which is one of those integrals a physicist may look up, rather than prove?). The expression for $\mu(T)$ given at the end of section 8.1.2 is obtained from equation (C.7) using $F(\varepsilon) = \int_0 g(\varepsilon) d\varepsilon$. And equation (8.9) for $U(T)$ is obtained with $F(\varepsilon) = \int_0 \varepsilon g(\varepsilon) d\varepsilon$.

Appendix D

Some useful constants

Boltzmann's constant	k_B	=	$1.38 \times 10^{-23}\,\mathrm{J\,K^{-1}}$
Avogadro's number	N_A	=	$6.02 \times 10^{23}\,\mathrm{mol^{-1}}$
Gas constant	R	=	$8.31\,\mathrm{J\,mol^{-1}\,K^{-1}}$
Planck's constant	h	=	$6.63 \times 10^{-34}\,\mathrm{J\,s}$
	$\hbar$	=	$h/2\pi = 1.05 \times 10^{-34}\,\mathrm{J\,s}$
Electronic charge	e	=	$1.60 \times 10^{-19}\,\mathrm{C}$
Speed of light	c	=	$3.00 \times 10^{8}\,\mathrm{m\,s^{-1}}$
Mass of electron	m	=	$9.11 \times 10^{-31}\,\mathrm{kg}$
Mass of proton	M_P	=	$1.67 \times 10^{-27}\,\mathrm{kg}$
Bohr magneton ($= e\hbar/2m$)	μ_B	=	$9.27 \times 10^{-24}\,\mathrm{J\,T^{-1}}$
Nuclear magneton ($= e\hbar/2M_P$)	μ_N	=	$5.05 \times 10^{-27}\,\mathrm{J\,T^{-1}}$
Atmospheric pressure		=	1.01×10^{5} Pa (i.e. $\mathrm{N\,m^{-2}}$)
Molar volume of ideal gas at STP		=	$22.4 \times 10^{-3}\,\mathrm{m^3}$
Permeability of free space	μ_0	=	$4\pi \times 10^{-7}\,\mathrm{H\,m^{-1}}$
Gravitational constant	G	=	$6.67 \times 10^{-11}\,\mathrm{N\,m^2\,kg^{-2}}$
Mass of Sun		=	$2.0 \times 10^{30}\,\mathrm{kg}$

Appendix E

Exercises

Chapter 1

1 Consider a model thermodynamic assembly in which the allowed (non-degenerate) one-particle states have energies $0, \varepsilon, 2\varepsilon, 3\varepsilon, 4\varepsilon, \ldots$. The assembly has four distinguishable (localized) particles and a total energy of 6ε. Identify the nine possible distributions, evaluate Ω and work out the *average* distribution of the four particles in the energy states. (See also Chapter 5, question 1.)

Chapter 2

1 Verify (2.28), (a) by working through the outline derivation given in the text, and (b) by using it to derive an expression for S $[= -(\partial F/\partial T)_{V,N}]$ which is the same as that obtained by method 1 of section 2.5.

Chapter 3

1 Below what temperature will there be deviations of greater than 5% from Curie's law (3.10), even in an ideal spin-$\frac{1}{2}$ solid?

2 The magnetization of Pt nuclei (spin $\frac{1}{2}$) is commonly used as a thermometric quantity at very low temperatures. The measuring field is 10 mT, and the value of μ for Pt is 0.60 μ_N. Estimate the useful range for the thermometer, assuming, (a) that a magnetization of less than 10^{-4} of the maximum cannot be reliably measured, and (b) that deviations from Curie's law of greater than 5% (see question 1) are unacceptable. In practice NMR techniques are used to single out the energy splitting; what is the NMR frequency in this case?

3 Negative temperatures? Show that for the spin-$\frac{1}{2}$ solid with thermal energy $U(\text{th})$, the temperature is given by $1/T = (k_B/\varepsilon)\ \ln\{[N_\varepsilon - U(\text{th})]/\text{U(th)}\}$. Hence show that negative temperatures would be reached if $U(\text{th}) > \text{N}\varepsilon/2$.

(a) A mad scientist, bowled over with the logic of (3.7), suggests that if should be possible to achieve negative temperatures in the spin system by an adiabatic reversal of the applied field (i.e. demagnetization and remagnetization with B reversed). Explain why this method will not work. (*Hints:* Plot temperature versus time for the process; and Fig. 3.7 should help also.)

(b) A negative temperature in a spin system can nevertheless be reached; indeed it is a prerequisite for a laser or maser. Look up and write a brief account of the methods used.

(c) Explain why negative temperatures cannot be contemplated in the assembly of harmonic oscillators.

4 The energy levels of a localized particle are $0, \varepsilon, 2\varepsilon$. The middle level is doubly degenerate (i.e. two quantum states have this same energy) and the other levels are singly degenerate. Write down and simplify the partition function. Hence compare the thermal properties (U, C, S) an assembly of these particles with the properties of the spin-$\frac{1}{2}$ solid.

5 Explain why iron is not used as the coolant for adiabatic demagnetization.

Chapter 4

1 Consider waves trapped in, (a) a one-dimensional, and (b) a two-dimensional box. In each case, derive $g(k)\delta k$ and compare your results with (4.4). Find the energy dependence of the density of states in ε for a gas of particles of mass M, comparing with (4.9).

Chapter 5

1 Repeat question 1 of Chapter 1 for a model assembly of four particles with the same energy states and with $U = 6\varepsilon$ as before, for the cases when the particles are, (a) gaseous bosons, and (b) gaseous fermions. Compare the results for the three cases.

2 (Not an easy problem.) As a game – which has no physical basis – work out the statistics of a gas obeying 'intermediate statistics'. The one-particle states of the gas may be occupied by $0, 1, 2, \ldots p-1, p$ particles, so that the two physical cases are $p = 1$ (FD) and p infinite (BE). Obtain expressions for $t\{(n_i)\}$ and for the thermal distribution, valid for any specified value of p. Check that the two limits give the correct results. (*Hint:* If you get stuck, see if your library has *Molecular Physics*, vol. 5, p. 525 (1962) in which all is revealed!)

Chapter 6

1 Check the derivation of (6.8) from (6.7), including the value of the constant C.

2 Using the integrals of Appendix C, verify the stated expressions for the mean and the rms speeds in section 6.2.

3 What is the distribution function in v_x for an MB gas? This is defined such that $n(v_x)\mathrm{d}v_x$ is the number of particles with x-component of velocity between v_x and $v_x + \mathrm{d}v_x$.

4 In an experiment to study the MB distribution, gas molecules are observed after they have escaped through a small hole in the oven. Show that the rms speed of the escaped molecules equals $2v_T$ (i.e. higher than the rms speed inside the oven). (*Hint:* Do question 3 first.)

5 Find the constant in equation (6.10).

Chapter 7

1 Calculate the percentage contribution of vibration to the heat capacity of O_2 gas at room temperature (293 K). The characteristic (Einstein) temperature for vibration in O_2 is 2200 K.

2 Work out the characteristic temperatures of rotation (7.7) for, (a) O_2, (b) H_2, (c) D_2, (d) HD. The masses of H, D and O are respectively 1, 2 and 16 times M_P. The internuclear distance in O_2 is 1.20×10^{-10} m, and in the hydrogen gases is 0.75×10^{-10} m.

3 Consider the properties of D_2. (a) Show that the A spin states (i.e. para-deuterium) are associated with odd-l rotational states, and the S spin states (ortho-deuterium) with even-l rotational states. (b) What is the ortho:para ratio in deuterium at room temperature? (c) What is the equilibrium composition at low temperatures?

4 Suppose the partition function Z of an MB gas depends on V and T as $Z \propto V^x T^y$. Find P and C_V.

Chapter 8

1 Check the derivation of (8.5). Hence work out the Fermi energy and Fermi temperature of the conduction electrons in metallic copper. You may assume that the conduction electrons behave as an ideal gas of free electrons, one electron per atom. The molar volume of copper is 7 cm^3. Show from a calculation of the Fermi velocity of the electrons (defined by $\hbar k_F = m v_F$) that relativistic corrections are not important.

2 Calculate the Fermi temperature for liquid ^{3}He, assuming the ideal gas model. The molar volume of liquid ^{3}He is 35 cm^3.

3 Verify (8.8), that the average energy per particle in an ideal FD gas is $\frac{3}{5}\mu$. Prove in addition that the average speed of the gas particles (at $T = 0$) is $\frac{3}{4}v_F$, where the Fermi velocity v_F is defined by $m v_F = \hbar k_F$.

4 A semiconductor can often be modelled as an ideal FD gas, but with an 'effective mass' of the carriers differing from the free electron mass m. A particular sample of InSb has a carrier density of 10^{22} m^{-3}, with effective mass 0.014 m. You may assume that the carrier density does not vary significantly with temperature (more or less valid for this material). (a) Find $\mu(0)$ and T_F. (b) Show that at room temperature (say 300 K) MB statistics may just be safely used. (c) Estimate μ at 300 K. (*Hint:* It will be negative – why? Note that the power series method of Appendix C cannot be applied accurately here.) How does μ vary with T above 300 K?

Chapter 9

1 Work through the steps leading to (9.3) and (9.4).

2 Estimate the condensation temperature T_B for an ideal BE gas of the same density of ^{4}He atoms as occurs in liquid ^{4}He (molar volume 27 cm^3).

3 The Stefan–Boltzmann law states that the energy flux radiated from a black body at temperature T is σT^4 J m^{-2} s^{-1}. Derive this law from first principles, assuming that the radiation may be described as a photon gas. (a) Explain the origin of the T^4 factor. (b) Compare your theoretical value of σ with the experimental value, $\sigma = 5.67 \times 10^{-8}\,\mathrm{W\,m^{-2}\,K^{-4}}$.

4 A vacuum insulated flask has the shape of a sphere (with a narrow neck!) of volume $5 \times 10^{-3}\,\mathrm{m^3}$. The flask is filled with liquid nitrogen (boiling point 77 K, latent heat $1.70 \times 10^8\,\mathrm{J\,m^{-3}}$). (a) Estimate the hold time of the flask (important practically if the flask is used as a 'cold trap'), assuming that the outer surface of the flask is at 300 K and that both surfaces behave as perfect black bodies. (b) How in practice is such a (Thermos) flask made more efficient? Explain.

5 Wien's law for black-body radiation (used for thermometry by colour) states that $\lambda_{\max} T =$ constant, where $\lambda_{\max}$ refers to the maximum in $u(\lambda)$, the energy density per unit wavelength. The experimental value of Wien's constant is about 0.0029 K m. Calculate the theoretical value of the constant as follows:

(a) Re-express Planck's law (9.12) in terms of wavelength λ instead of frequency ν. Explain why the peaks in $u(\nu)$ and in $u(\lambda)$ do not occur at the same photon states.

(b) Using the usual type of dimensionless variable $y = ch/\lambda k_B T$ obtain an equation for $y_{\max}$, the y-value corresponding to $\lambda_{\max}$.

(c) Solve the equation numerically or graphically. (*Hint:* the solution occurs at y just less than 5.) Hence compute Wien's constant.

(d) Show that the maximum in (9.12) occurs at a y-value just less than 3, as mentioned in the text.

6 Starting from (9.14) for the thermal energy of a solid: (a) Show that at high temperatures, the classical result $C_V = 3Nk_B$ is recovered. (*Hint:* only small values of y are involved, so expand the exponential.) (b) Show that the Debye T^3 follows at low temperatures. (c) At even lower temperatures, (9.14) will become in valid because it is based on the density of states approximation. Estimate the temperature below which (9.14) will severely overestimate U (and hence C_V) for a sample of size, (i) 1 cm, and (ii) 1 μm. The speed of sound in a typical solid is $4000\,\mathrm{m\,s^{-1}}$.

Chapter 10

1 (a) Verify the derivations of (10.4) and (10.5).

(b) (Harder) Find the analogous expression for S for 'intermediate statistics' (see question 2 for Chapter 5), and check that it has the correct limits.

2 Consider the formation of case 2 vacancies (often called Frenkel defects) as introduced in section 10.3. The defect occurs when an atom leaves a normal site and moves to an interstitial site. Suppose the crystal has N normal sites and N_1 interstitial sites (these numbers will be the same within a small multiple which depends on the crystal structure). The energy of formation of the defect is Φ. Following the method of section 10.3, show that the number n of vacancies at temperature T is

$$n = (NN_1)^{1/2} \exp(-\Phi/2k_B T)$$

(*Hint:* Disorder now occurs in the arrangement both of the vacancies and of the interstitial atoms.)

Chapter 11

1 Estimate the strength of the effective interaction field (11.2) in a strong ferromagnet such as iron ($T_C = 1043$ K). You may assume that iron is a spin-$\frac{1}{2}$ solid, with $\mu = \mu_B$. This is not accurate, but it will give an order of magnitude.

2 Estimate the typical ordering temperature for a nuclear spin-$\frac{1}{2}$ solid. Assume that all the nuclei have a moment μ_N, that they are separated by 0.3 nm and that the interactions arise from the magnetic dipole influence of about eight nearest neighbours. (Don't try to be too accurate!)

3 In the case of beta-brass (section 11.4) show that, in the mean field approximation, the structural contribution to the internal energy may be written as

$$U = U_0 - 2Nm^2V$$

where $U_0 = 2N(V_{\text{CuCu}} + V_{\text{ZnZn}} + 2V_{\text{CuZn}})$, and $V = V_{\text{CuCu}} + V_{\text{ZnZn}} - 2V_{\text{CuZn}}$. Hence derive an expression for the ordering temperature in terms of the bond energy difference V. (*Hint:* Work out the number of each type of bond as a function of the order parameter m, assuming that the occupation of each site is uncorrelated with its neighbours – the mean field assumption.)

Chapter 12

1 Verify that (12.4) leads to the FD distribution, as stated in section 12.1.2.

2 (a) If the atoms were interacting with a gas consisting of real (massive) bosons, what equations should replace (12.4) in order for the BE distribution to follow?
(b) Repeat (a) for an MB gas leading to the MB distribution.

3 Verify that $Z_A = Z^N$, for an assembly of N localized particles, using the multinomial theorem method. (See note 5 after (12.7).)

Chapter 13

1 Work through the derivations of (13.5) and (13.6), the basic ideas in the use of the grand canonical ensemble.

2 As outlined in section 13.3.3, use (13.13) to derive the heat capacity C_V of an ideal gas which is (i) monatomic, (ii) diatomic. Express the result in terms of the N rather than μ (compare (13.14)).

3 Consider the chemical reaction $2H_2 + O_2 \rightleftharpoons 2H_2O$. Writing subscripts 1 for H_2, 2 for O_2 and 3 for H_2O, show that (i) the corresponding result to (13.22) is $2\mu_1 + \mu_2 = 2\mu_3$ and (ii) the law of mass action (compare (13.24)) for the reaction is $N_1^2N_2/N_3^2 = K(V,T)$.

4 A particular chemical reaction doubles its rate for a 10 K temperature rise near room temperature (say from 290 to 300 K). Estimate the activation energy for the reaction.

Chapter 14

1 Describe how Fig. 14.1a and 14.1b are related. How does the electron velocity vary over the same range?
2 Verify, as stated in the text between (14.2) and (14.3), that the density of states $g(\mu)$ is given by (8.3) but with the mass changed.

Chapter 15

1 Sketch the expected entropy behaviour that you would expect across a first-order, a second-order and a third-order phase transition (see section 11.1). How would you classify the following?
  (a) A superconducting transition in zero applied magnetic field (see Fig. 15.1).
  (b) A superconductor where the transition temperature has been suppressed by the application of a modest magnetic field (see Fig. 15.1 again, and guess what will happen!).
  (c) The superfluid transition in ^{4}He (see sections 14.3 and 9.2, and do not expect a tidy answer).

2 Verify the calculations made following (15.1).
3 Work through the steps leading to (15.3), the pressure of a relativistic FD gas.
4 Verify the numerical values given for the constants K (equation (15.5)), and K_1 (equation (15.6)), and hence check the upper mass limits for white dwarf and neutron stars.
5 Explore the question, discussed before (15.7), as to whether the relativistic speed correction is ever important for the stability of a neutron star.

Appendix F

Answers to exercises

Chapter 1

1 $\Omega = 84$; average distribution (1.33, 1.00, 0.71, 0.48, 0.29, 0.15, 0.05, 0, 0 ...) is almost exponential, as for a large assembly (Chapter 2).

Chapter 3

1 $2.6\,\mu B/k_B$.
2 $5.7\,\mu$K to 22 mK; 91 kHz.
4 Note that Z is a perfect square. The limiting value of S is $Nk_B \ln 4$.

Chapter 5

1 (a) 9; (1.33, 1.00, 0.78, 0.44, 0.22, 0.11, 0.11, 0,0, ...)
(b) 1; (1, 1, 1, 1, 0, 0, 0...).
2 $n_i/g_i = 1/\{\exp[-(\alpha + \beta\varepsilon_i)] - 1\}$
$- (p+1)/\{\exp[-(p+1)(\alpha + \beta\varepsilon_i)] - 1\}$

Chapter 7

1 1.24%.
2 (a) 2.1 K, (b) 85 K, (c) 43 K, (d) 64 K.
3 (b) 2, (c) pure ortho-deuterium.
4 $xNk_BT/V, yNk_B$.

Chapter 8

1 7.1 eV, 82 000 K, $1.7 \times 10^6\,\mathrm{m\,s^{-1}}$.
2 5.1 K.
4 (a) 139 K, 0.012 eV, (c) $\mu/k_B \approx -225$ K.

Chapter 9

2 3.2 K.
4 (a) 3.6 h. (b) include silvering and superinsulation in your answer.
5 $y_{max} = 4.9651\cdots$; max in $u(\nu)$ at $y = 2.8214\cdots$.
6 (c) 20 μK, 0.2 K.

Chapter 11

1 1500 T.
2 About 50 nK.

Chapter 13

4 0.52 eV.

Index